GW01459263

RADIO
A Guide to Broadcasting Techniques

RADIO

A Guide to
Broadcasting Techniques

ELWYN EVANS

Formerly Head of the Radio Training Section, BBC

BARRIE & JENKINS
COMMUNICA - EUROPA

© Elwyn Evans 1977
First published in 1977 by
Barrie and Jenkins Ltd
24 Highbury Crescent London N5 1RX

All rights reserved. No part of this
publication may be reproduced in any
form or by any means without the prior
permission of Barrie and Jenkins Ltd.

ISBN 0 214 20378 6

Printed in Great Britain by The Anchor Press Ltd
and bound by Wm Brendon & Son Ltd
both of Tiptree, Essex

For Margot

CONTENTS

CONTENTS

PREFACE

The first radio programmes went on the air soon after the First World War, and radio as an institution and an art has been with us ever since. Faced with competition from television it may have shifted its ground here and there, but it shows little sign of surrendering its grip on the mass audience. In the Third World it probably provides more men, women and children with information, education and entertainment than all the other media put together.

In writing this book I have had the needs of three groups of people particularly in mind: those who want to break into radio as contributors; programme staffs, especially in developing countries, and students of mass communications.

For most of the ideas expressed I claim no credit. They are the common property of broadcasters in Britain and the Commonwealth. My aim has been to put the insights of a multitude of specialists down on paper and turn them into a coherent body of doctrine, checked against my own experience, which happens to have been unusually wide.

The book is intended to be read straight through. But I dare say some people will want to consult particular chapters only; so a few basic points have been allowed to appear in more than one context.

The advice given within is put rather dogmatically at times. May it please be remembered that when I write 'Don't do so-and-so' what I mean is 'Most practitioners agree that in most cases, though not of necessity in all, so-and-so is not an effective course to follow.' The imperative formulation is shorter.

ACKNOWLEDGEMENTS

The list of those to whom I owe thanks for ideas and information is endless. It ranges from freelance actors in Wales to producers in West Africa and speakers at BBC Staff Training seminars in London. I do thank them all, but particularly those I met during my stint in Training – the colleagues and visiting experts (some are named in the text) who extended my horizons, and the students, from many parts of the world, whose questions forced my notions into sharper focus. Two colleagues, Ronnie Morgan and his assistant Mike Grimshaw, added immeasurably to my knowledge of radio's technical side.

They have also read the technical parts of this book in manu-script. Ted King, of the British Forces Broadcasting Service, checked the section on disc-jockeying, and Richard Keen went through every chapter. As a result of these kindnesses various gaffes have been eliminated. Any that remain are without doubt 'the responsibility of the author'.

For permission, most generously accorded, to reproduce extracts from broadcast scripts I am indebted to the BBC (in respect of scripts by Ian McDougall, Nesta Pain and myself) and to Vernon Bartlett, René Cutforth, Don Haworth, Thea Holme, Prof. Peter Murray, Mrs W. R. Rodgers and Gwyn Thomas. I would also thank the editors and authors who allowed me to quote from newspapers and magazines (named in the text).

Finally, may I say how grateful I am to Patricia Parkin, formerly of Barrie and Jenkins, for her constructive interest in the book and to David Thomson and the Rev. E. H. Robertson for many favours, but above all their encouragement.

E.M.E.

SOME TERMS

Cubicle. BBC name for the room (next to the studio) in which a production is directed. The room contains the console, disc and tape machines, a loudspeaker (or two loudspeakers in the case of stereo productions), etc. A window above the console enables the producer and his technical assistants to see what the performers are doing. Other names for the cubicle: Listening Room, Control Room (but the BBC uses 'Control Room' in quite a different sense).

Leader. Short for Leader Tape. A leader is patched in to indicate the precise point at which a programme, or insert, begins.

Light cue. Also called a Flick, a Light, or a Green. The signal given to an actor or other performer to speak a line after a period of studio silence (e.g. during the playing of a recorded sound-effect).

Panel. Also called a Console or a Desk. The piece of equipment into which the outputs of all the sound sources in a production (microphones, disc and tape machines, etc.) are fed, there to be 'mixed' by a senior studio manager (the Panel Operator), under the direction of the producer. Panels are used in Outside Broadcasts as well as studio productions.

Presentation. Usually means the work done by staff announcers (see Chapter 5). But it can also mean the work done by a particular sort of magazine compere (see Chapter 8).

Producer. The classical radio name for the person who takes responsibility for a programme at all its stages. The TV distinction between producer and director hardly exists in radio (though attempts are being made to introduce it).

Studio Manager. A producer's technical assistant. Not a very good term: for one thing, *studio* managers are often employed on OBs. As the invaluable human bridges between the artistic and the engineering sides of radio they have been the subject of innumerable administrative takeover bids. In consequence they have been called at various times and by various organisations Balance and Control staff, Programme Engineers, Technical Operators, Programme Operators, etc. etc.

Talk-Back. Device enabling a studio producer, during a rehearsal, to address his cast without leaving the cubicle. Under transmission, or pre-recording, conditions, it enables a studio producer to address any performer, or an OB producer to address any commentator, who is wearing headphones. *Reverse* Talk-Back enables, e.g., a compere or disc-jockey to address the producer in the cubicle while a disc or tape is being transmitted.

1: TALKING TO AND TALKING AT

One speaker at one microphone: a single voice addressing the audience direct. Let's start with this basic form of communication by radio.

Compared with television, radio is a non-compulsive medium. When someone talks to us we normally expect not only to hear him but to see him. If all we get is a voice coming out of a box, the voice of a permanently invisible speaker, we are in a highly artificial situation. In radio the link between speaker and audience is frail. Where there is no vision people get bored that much quicker. This is a point that many radio practitioners prefer to overlook; yet it's fundamental to any honest consideration of the medium. A listener is a person willing, if not anxious, to get away. It's the job of a speaker to keep him hooked.

How is this to be done? How can his interest be first seized and then held? Occasionally the importance or novelty of the message is enough. Some items of news, some eye-witness stories, even some weather forecasts (in Britain at least) will command attention however badly they are put together and put across. But most communicators aren't so fortunate in their subject-matter. These people have to depend on the style of their message, the way it's conveyed.

It has been proved over and over again that the most effective speaker is the personal speaker. He may be reading a script, but he *sounds* as though he's talking to me alone. My conscious mind may be aware that he isn't doing anything of the sort – but, as in the theatre, it's the subconscious impression that counts. If a radio speaker, thanks to the way his script is written, makes me *feel* he's talking to me personally, it becomes much harder to switch him off.

For most of us there is a fearful gap between the Spoken and the Written Word. To write in the way we talk seems almost a contradiction in terms. Certainly it's a reversal of what we learnt

15

at school. Nevertheless, this is what you have to do when you prepare a script (within limits, of course, i.e. minus the hesitations, ellipses and ambiguous syntax of ordinary life). And when you give your talk at the microphone you must attempt something harder still. You mustn't be content with reading out what you've written: you must converse, really converse, with someone who isn't there.

WHY SCRIPTS?

Let's concentrate for a while on the script. But before doing so we might ask why we have scripts at all. Why go through the elaborate charade of first putting words on paper and then pretending we're making them up as we go along? Why not *actually* make them up as we go along, like disc-jockeys? The fact is that most people *need* scripts if they are to talk solo for more than a few seconds at a time. Otherwise they repeat themselves, forget vital points, over-run or dry up. You may say this doesn't matter if the speaker is being pre-recorded: he can always have another go. True, but tape-editing is a time-consuming and expensive process.

At one time whole families used to gather round 'the wireless'. They still, to some extent, gather round the television set, but practically everybody, from Fulani cattle-herder to Birmingham housewife, now listens on a set of his or her own. In radio the audience to be aimed at is *an audience of one* (infinitely repeated). And as useful a way as any of drafting a script is to *imagine* that audience of one, pretend you're explaining a point or telling a story to him or her, and then put down on paper what you've said – not immediately, but after you've tried to phrase your message in different ways and settled on the best. Some broadcasters, including me, find it helpful to mutter the words aloud.

The result can be something as effective as this (the opening of a talk called 'Life Begins at Seventy', by Vernon Bartlett):

I can't claim any special merit for the fact that I've survived the hazards of this world for just on seventy years. Quite a lot of people have done that in the past. Quite a lot will do it in the future. Why then am I presuming to talk to you about it today? Mainly because I happened to broadcast a talk ten years ago in which I claimed that life began at sixty. I was then on the point of pulling up my roots and going off to live in Malaya, and not very many people decide to

16

emigrate at so advanced an age. And, in a way I certainly hadn't anticipated when I prepared the broadcast, my claim that life began at sixty turned out to be true.

I came to the studio direct from hospital and as soon as the talk was over I went back to hospital, where they had to chop me about so vigorously that I wasn't really expected to survive. I did survive, after all – here I am, talking to you today.

But, you may object, Vernon Bartlett's *subject-matter* was personal: you can't deal with *every* topic in a personal way. To this there are several answers. The short answer is, if you can't speak personally on the air you had better not speak at all, because no one's going to listen to you. Of course every medium has its limitations; there are some subjects that defy radio treatment, subjects for which spoken words are just not enough – advanced chemistry, for instance, with its diagrams and equations, or, on a homelier level, how to tie knots. But if you believe that a subject is resistant to radio why broadcast a radio talk on it? You should go on television or write a book. There's no percentage in trying to solve the difficulty by composing your talk in a non-radio (non-conversational and non-personal) style.

I do assure you, however, that the number of subjects you can't put over personally and conversationally is much smaller than people think: nine times out of ten the trouble lies not with the subject but with the speaker, and his lack of imagination. As a talks producer I found that top experts, whether on fly-fishing or philosophy, could almost always communicate with the rest of us: it was the chap who wasn't quite so good who wallowed in a Sargasso Sea of impersonal jargon, making no contact with the listener.

Lastly – and this is vital – it should be remembered that 'personal' doesn't mean twee, or cloying, and that 'conversational' doesn't mean subliterate. There's no reason why a don should sound like a disc-jockey. But there's all the difference in the world between a don when he's lecturing – talking *at*, or rather reading at, a whole crowd – and the same don conducting a tutorial with an intelligent pupil and talking *to* him. All the don has to do, essentially, when he broadcasts is write his script à la tutorial and not à la lecture. The process really couldn't be simpler. It's all a matter of attitude and approach.

One main reason why the old BBC Third Programme had to be killed off, in spite of its great achievements in features and drama, was its policy for talks. Owing to the rather touching intellectual snobbery of its founding fathers the convention was soon established that producers should not try to improve the offerings of speakers above the rank of senior lecturer. Hence all those turgid discourses by mandarin figures who hadn't bothered to learn the elements of broadcasting. Long before the end the Third Programme had become a home for anti-radio, and listeners had deserted it in droves. Every now and again, however, a speaker would come along who could not only think but communicate. What could be harder than to describe in words the principle of the arch and the dome? Yet this is what Peter Murray managed to do – and in a way that riveted one's attention. Here he is, talking about the huge dome of Florence Cathedral, and how Brunelleschi managed to get it up.

If you have ever watched an arched opening being built you will know that the arch, like the wheel, is one of the greatest of human inventions. You build an arch by making a wooden framework, exactly the size and shape you want, putting this frame in the wall at the point from which the arch is to spring and then laying your bricks, or stones carved in wedge shapes, on the framework. Obviously, if the framework were to be removed, or fell down, all the stones would come crashing down as well; but if you fit all the stones together, inserting a wedge-shaped one – the keystone – at the very top, then, when the arch is complete, you can take the wooden frame away and all the stones, under the influence of gravity, will promptly try to fall down. Because the ring is complete and because the keystone is wedge-shaped, what in fact happens is that the stones all jam together solidly to form an arch which can carry a considerable load: indeed, within limits, the greater the load on it the more firmly the arch is bound together.

A dome is no more than an arch revolved through 360 degrees, so if you can build an arch on timber centring, as it is called, then you can also build a dome. The snag that I mentioned comes from the limitation imposed by the wooden frame.

And he went on to explain, in memorably interesting detail, exactly what the snag was and how Brunelleschi overcame it.

You can imagine how the average expert would have dealt with the subject:

The theory of the arch may best be approached by reference to its mode of construction. The primary stage in the erection of an arch is

18

the construction of a wooden framework of precisely the dimensions required. This is placed in the wall at the point from which the arch is required to rise, bricks or wedge-shaped stones being then laid upon the framework itself. It might be anticipated that on the removal or collapse of the framework the bricks or stones would in turn fall to the ground. However . . .

It doesn't bear thinking of.

THE DIFFICULTIES OF BEING A LISTENER

When you talk to someone face to face he indicates in a hundred ways whether he's following you or not and whether he's interested or not in what you're saying. Unless you're a complete bore you take note of these unconscious signals and adapt your conversation accordingly. On the air you have no such help. I've suggested already that in writing a script you should have an imaginary listener in mind and make sure, phrase by phrase, that you sound as though you're *talking* to him. But this isn't enough. You should also ask yourself, point by point, 'Am I still *interesting* him?' and 'Am I still making myself clear?'

Listening to the radio is a single-sense activity, like reading. But a reader can do so much that a listener can't. Readers can anticipate: if the first words of an article don't appeal they can glance down the column to see if it gets more interesting. But a listener can't tell what's coming next. If he's put off by a feeble opening the odds are that he won't give the rest of the broadcast a chance: he'll switch off right away. Then again, a listener can't skip. Even if you capture his attention at the start, you still have to hold it, moment by moment. Once his mind starts to wander you've lost him. What this means is that whereas authors can afford a few dull passages, provided the piece as a whole is interesting, radio speakers can't. They have to be interesting *all the time*. A reader can choose his own pace but a listener is bound to the pace of the speaker. Normally this is too fast for us to take in complicated figures and detailed information, just as it's too slow for us not to be irritated by clichés, mixed metaphors and other stylistic deficiencies that a reader may hardly notice. A personal style doesn't mean a careless style: on the contrary.

Finally, a listener can't stop and think, and he certainly can't refer back. If he tries to do either he instantly ceases to hear what's being said *now*. So when you start writing remember that clarity and logic are of the first importance. Don't confuse and irritate your listener with inconsistencies, holes in the argument, irrelevancies and loose ends. Having made two points and started on a third, don't suddenly go back again to point one. 'Radio,' Laurence Gilliam once remarked, 'is the art of communicating meaning at first hearing.' Not an exhaustive definition, but one worth painting in letters of gold on the walls of every talks studio in the world.

AT THE MICROPHONE

A good script doesn't guarantee a good broadcast, but it takes you three-quarters of the way. Having composed the script in conversational style, all that remains is to utter it in conversational style. This isn't difficult as long as you don't commit any of those follies which keep talks producers in full employment. Don't raise your voice at the end of a sentence; don't adopt an artificial tone or a special accent. Don't speak any louder or quieter, any lower or higher, any slower or quicker, than usual. In short, don't put on any sort of act: your ordinary way of speaking is perfectly all right: if it weren't you wouldn't have been asked to broadcast.

But this is negative advice. The essential, positive, point is that you must maintain the right *attitude*: as in composing the script so now in the studio. The listener, we agree, needs to feel that he's being spoken to personally. This can only happen if the broadcaster feels that he himself is talking personally, to a particular individual. There are at least two ways of achieving this necessary illusion. You can imagine that the person you had in mind when you wrote the script is now listening. Or you can personify the microphone. This is what many broadcasters (including me) do: we keep looking at the microphone, gesticulating towards it, *addressing* it.

At this stage a lapse of memory comes in very handy. During the actual recording, or transmission, you must forget the effort it cost you to put the script together; you must forget that you've just been rehearsed; you must will yourself to believe that 'it's all happening now', that you're saying it all for the very first time, and meaning what you say.

20

If you find you can't do this because you're anchored to a script, try paying out some more cable. Give yourself some latitude. Alter the occasional phrase as you go along, while sticking to the text as a whole. The new phraseology may not be as well-chosen as the old and you may find yourself stumbling and hesitating a little, but you will recover some measure of spontaneity. And this will give the broadcast a psychological lift. After all, the script is only a *reminder* of what you want to say. No prizes are offered for verbal accuracy.

A final point, of importance to unpractised speakers. Don't get tangled up in your punctuation. The principles according to which we group words on the page are vastly different from those we follow in speech. Consider a very ordinary sentence:

If, on the other hand, Spurs *don't* manage to beat Arsenal tomorrow, then, whatever they may do on Saturday against Liverpool, they can bid farewell to any hope they may have had of winning the Cup.

Spoken, the words would be grouped very differently: rather like this, perhaps:

If on the other hand Spurs *don't*/manage to beat Arsenal tomorrow/ then whatever they may do on Saturday/against Liverpool/they can bid farewell/to any hope they may have had/of winning the Cup.

When you speak, *never try to observe the punctuation*. In fact, if you have not been trained to read aloud, the best thing you can do is, within reason, to ignore the punctuation and concentrate for all you're worth on the meaning. Take care of the sense and the sounds will take care of themselves.

POSTSCRIPT FOR JOURNALISTS

In spite of the example set by Vernon Bartlett, Alistair Cooke and others, most newspaper men who transfer to radio find it difficult to accept, and still more difficult to practise, the principles stated above. I implore them to make an effort to realise that the very fact that their words will be heard, not read, *fundamentally* alters the terms of the writing equation – and, having realised it, to draw the appropriate conclusions.

21

POSTSCRIPT FOR PARSONS

The quasi-personal relationship between speaker and listener must be established delicately, by implication, and certainly without over-emphasis. The parsonical or advertising hard-sell – 'Have *you* ever thought . . .?', 'What are *you* doing . . .?' – is off-putting in the extreme.

2: DRAFTING A SCRIPTED PIECE: SOME RULES OF THUMB

Now for some bits of practical advice, based on the proposition that a talk should be conversational, interesting and clear. (I'm using 'talk' in its widest possible sense, which ranges from reports in a news bulletin or items in a magazine to full-scale solo broadcasts lasting fifteen or twenty minutes.)

OPENINGS

There are several tried and tested methods of starting effectively. One is the Big Bang technique – saying something that will make listeners sit up. Thus:

Syria, on my count, has had 22 regimes in the past 28 years. Changing from one to another has sometimes proved a bloody affair.

(Ivor Jones)

A great silent cheer went up from the South last week.

(Alistair Cooke)

The Senate doves are getting restive again.

(Charles Wheeler)

A second method is to lay a false trail – an ancient trick, much practised by English essayists as long as essays continued to be written. You start with an attractive notion that seems to have no relevance to the advertised subject, which you work round to by degrees. This is too leisurely a device to be used often, but it can come off extremely well. When Willy Brandt took over the Chancellorship of West Germany this is how Ian McDougall began a talk on the changes to be expected in German foreign policy:

All national images die hard. You have only to study newspaper cartoons the world over to realise that Englishmen wear bowler hats and thick moustaches, Frenchmen wear berets and thin moustaches, Americans wear dark glasses and Hawaii shirts and smoke cigars.

Germans are the easiest targets of all, clothed in jack-boots or leather shorts, their hair cropped to the skull, their women looking like something out of a Wagner chorus. This is all quite natural because the human mind has to visualise at least something about another nationality. It would be dull and unsatisfactory to accept the modern truth, which is that most Europeans, especially the young and middle-aged ones, now look and behave remarkably alike. You'd have to be pretty observant to distinguish between a crowd of Germans and a crowd of Britons or Frenchmen, if you couldn't also identify their immediate surroundings, but the fact remains that we all do like to distinguish.

What I'm really coming to is that Herr Brandt's assumption of the Chancellorship may well have marginally shifted the stereotype image of Germany traditionally held abroad.

Again, you can make an opening point and then leave it, or half-leave it, for a while. This is René Cutforth:

At the beginning of June 1944 I'd spent almost exactly two years as a Prisoner of War in Europe.

There are four main stages in a Prisoner of War's progress. There's the first stage of release from the tension of battle, when the new prisoner talks all day long at top speed and has nightmares all night long. And after that guilt and resentment take over, and he spends six months plotting to escape. And after that he plays ping-pong and learns Russian and hates the sight of everybody. And after that he does nothing at all. He lies about; he doesn't even read. There's no point in moving from A to B because not only is B just as unpleasant as A, it is also exactly the same as A. You know if you repeat a word, your own name for instance, often enough it becomes quite meaningless. Well, in a Prisoner of War Camp after two years, everything, every squalid thing, has been repeated over and over again until there's no meaning in anything, and then you begin to forget how to talk because there's nothing to be said. You're then, as they say, round the bend, and that's the time when you may very easily fold up and die, usually of pneumonia.

After that first sentence, however long Cutforth talked about camp conditions in general and how the average man reacted to them, one felt a need to keep on listening until he came back to the point and told us what had happened to him personally. The opening set up an expectation in the mind which simply had to be satisfied. Sometimes, as in Alistair Cooke's broadcasts, a speaker doesn't revert to his original point until the very end – a fine way, in cunning hands, of creating, and maintaining suspense. Lastly (this isn't meant to be a complete list) you can

24

start with a concrete instance and let an argument grow out of it. I began a BBC talk in this way:

Even in our quiet suburb we get the occasional road accident, and one happened near us the other morning. A milk delivery-van, maximum speed ten miles an hour, wanted to cross the main road from one side street to another. There was a car approaching, but it was a long way off so the van started to cross over. Unfortunately the car was going at about sixty, in a built-up area. It couldn't stop, there was a rending crash, and milk bottles flew in all directions. My wife happened to be walking down the street – there was no one else around. So she went up to the old milkman, who was a bit shaken but not hurt, and said if he wanted a witness she was available. She gave her name and address and pushed off.

From this opening there developed some thoughts about the duties of a citizen in various situations. Generalisations should usually be arrived at in this way, and as far as possible you should let your listener work them out for himself. What you should never do is start the other way round, directing a cloud of abstractions, or hurling lumps of solid fact, at your listener before you've even got him interested in the subject. You are, let us say, working in a local radio station, and you've been asked to do a three-minute report on an Exhibition of your city's Historical Treasures. *Don't* write your piece on these lines:

(i) Where the Exhibition is being held: reference to the inaugural speech by the Lord Mayor: opening and closing hours: how long the Exhibition will run.
(ii) General nature of the Exhibition.
(iii) Examples of what may be seen.
(iv) Conclusion: everyone should attend.

It would be much better done thus:

(i) Yesterday I held in my hand the very slipper that Lady So-and-So wore at the dance where she met Prince Something. I also saw the famous letter written from the Tower of London by XYZ thirty-six hours before he was executed.
(ii) Where was all this?
(iii) What's the Exhibition about?

and so on.

It's possible, of course, for your opening to be *too* punchy, or punchy in the wrong convention:

On Christmas Day the lemons ran out in the pub. 'The Saloon and the Public Bar will have to do without,' said the Landlord . . .

25

A bright, attacking first paragraph according to the canons of the short story. But no one would ever start a conversation in this way, or a speech. For a radio talk it doesn't ring true.

Trained newspaper men need to be especially careful with their openings. In particular they must unlearn the technique of encapsulating a whole report in the first sentence. Compare the way three London papers began their accounts of the same event.

(a) A circular asking for heart donors has been sent to hospitals by Mr Donald Ross, leader of Britain's heart transplant team, Dr Geoffrey Spencer, head of the intensive care unit at St Thomas's Hospital, London, said yesterday.

(b) Patients near the verge of death at Britain's largest intensive care unit at St Thomas's Hospital, London, were 'surrounded by a gang of vultures waiting to snatch out any useful organ from the cornea to the heart', Dr Geoffrey Spencer, the unit's director, said yesterday.

(c) The biggest row so far over heart transplants and spare-part surgery broke yesterday. The doctor in charge of Britain's largest intensive care unit attacked 'vultures trying to snatch organs'.

Of these three openings only the last would do for radio. The second is clumsy: a quotation followed by 'AB said' is old-fashioned journalese: the colloquial equivalent is 'AB said that . . .' The first opening, a particularly bad example of the 'AB said' construction, is so clogged with information as to be almost unintelligible to a reader: a listener would make nothing of it. By contrast the third version (beginning 'The biggest row so far') seizes one's attention. It also happens to be written in good straightforward English.

A warning (and a highly practical one). Listeners often take a moment or two to get mentally tuned in. The one thing wrong with the Ivor Jones opening at the beginning of this section is that if someone had missed the first word he'd have missed everything.

KEEPING UP THE INTEREST

Having made a good start, don't throw away your advantage. Take care to ration your facts. Never try to pack in as many as you could in an article of the same length. Above all, avoid masses of figures. '£238,375,218' may convey something to a reader – though less than financial writers may suppose. Over the air the figure is literally meaningless. It takes five seconds to pronounce – try it – and by the time a listener reaches the word 'pounds' he will have forgotten how many millions there were. *Use round figures.*

Better still, get as far away from statistics as you possibly can, and talk in pictures, images, concrete terms. When I was once asked to describe the construction of an enormous new steel works the PR man supplied me with pages of information – length, height, width, number of windows, the lot. What I said on the air was:

This new building lies at the edge of the sea; it's as big as seven Canterbury Cathedrals and it's painted sand-yellow.

I then added a few supporting statistics, but all the impact lay in the original image.

Significant detail can sharpen a picture. 'We were informed that because of bad weather ahead we should have to make an emergency landing' is prosaic. 'The air hostess came round and told us that because of storms ahead we'd have to make an emergency landing in Lagos' takes longer but has considerably more punch.

You can't maintain interest if your talk isn't clear. So emphasise a few main points and don't worry about the minor ones. Ideas are not born free and equal. And do *signpost* your intellectual journey. Give the listener some notion of the direction you're going to take, and let him know when you're about to move from one point to the next. Rhetorical questions ('But where do Manchester United come in?'), recapitulations ('We agree, therefore . . .'), anticipatory phrases ('Let's turn to . . .'), occasional summings-up: all these can be very useful.

LANGUAGE

Clichés and moribund language are sure dispellers of interest. Equally off-putting is an elaborate convoluted style, with sentences full of subordinate clauses, and the principal subject well removed from the principal verb. Since, as we know, a listener can't stop and think, or refer back, the odds are he'll get hopelessly confused. Furthermore, since the periodic style is now a lost art, sentences of this kind are likely to be ungainly as well as unintelligible. On the other hand, don't imagine that in radio all sentences must be short – that idea is a mere superstition. The length of a sentence depends very largely on the punctuation. 'I came. I saw. I conquered.' is three sentences. 'I came; I saw; I conquered.' is one. So what? Whether your sentences look long or short *on the page* doesn't matter: the important thing is that you should arrange your words simply and straight-forwardly. The English language has a natural word-order. Why not use it?

If you want to sound colloquial you must go for concrete nouns, not abstract, and verbs in the active voice, not the passive. Here's an extract that exemplifies three sins. It's need-lessly complex, it contains strings of abstract nouns, and it's joined together by verbs in the passive:

A policy of demolition of non-slum housing in slum areas and its wholesale replacement by council estates is often pursued by local authorities regardless of the fact that modernisation of such buildings when familiar to and appreciated by their inhabitants is frequently supported by both economic and social arguments.

What it means, I suggest, is this:

Local authorities often knock down houses that are not slums at all simply because they're in slum areas. And they replace them wholesale by council estates. What these councils don't realise is that often there are both economic and social arguments for modernising buildings of this kind, if those who live in them know them and like them.

If that *is* what the speaker means, why doesn't he say so?

It's a safe general rule that you should avoid purely literary turns of phrase. Take this, from a travel talk on India:

Bell clanging, brakes grinding, wheels squealing on the tracks, the ancient tram successfully blocks my way.

We never use the 'participle absolute' in ordinary speech. On the air it merely advertises the fact that the speaker is reading a script. Again, from the same talk:

The Sikh driver gets his old car to accelerate. The gearbox protests, the bodywork rattles, but the dented black mudguards and the yellow roof seem to flash by, and are soon lost in a cloud of blue smoke from the exhaust.

28

What one would *say* is

... the dented black mudguards and the yellow roof seem to flash by, and they're soon lost ...

To omit personal pronouns as one does in writing is a great give-away. On the other hand, the preposition 'that' should be omitted as often as possible. 'He said that he thought that he'd go' is clumsy. 'He said he thought he'd go' is much better.

Unintended rhymes and puns always come between a speaker and his audience.

I went down the steps of the cellar and switched on the light. Barrels of mild on the left, bitter on the right.

One's ear would pick up the rhyme immediately.

The Marketing Board's losses are said to be due to a recession in demand for palm-oil kernels. But Colonel Harrison thinks that prospects for future trading are bright.

It reads well enough, but the unconscious play on words – kernels, Colonel – makes it *sound* ridiculous.

This may seem finicky. Let me quote in support a passage from Professor John Hilton's famous Talk on giving a talk.

I've fished out of the waste-paper basket my first shot at the sentence with which I opened. . . . Here it is. 'I've wondered for days what kind of thing to say to you in this last of my broadcasts.' No wonder that went into the waste-paper basket! 'Day', 'say', 'last', 'broadcasts'. Awful! If I'd used that I should indeed have missed the mark. With you, I mean. You wouldn't have known, perhaps, what was wrong, but you'd have had a muzzy sort of feeling. No; one's got to be on the job all the time cutting out sound-clashes and echoes.

Verbal repetition, on the other hand, which so many writers assiduously avoid, often doesn't matter in the least. Here's an extract from an unscripted discussion on Bertrand Russell, broadcast a year or two before he died:

Kee: It seems odd if a man of such undoubtedly enormous intellect should turn out to be so ineffectual in public affairs.
St John-Stevas: I don't find that odd at all. One would expect some-one who has made his main business philosophy to be ineffectual in public affairs. One wouldn't expect him in fact even to *want* to take part in public affairs.

If the verbal repetition *sounds* all right it *is* all right. If it doesn't, it isn't – thus:

In his view the Premier's view was far too complacent. He demanded that action be taken with a view to remedying the situation.

Those three 'views' defeated experienced newsreaders.

When writing out your piece, remember to elide as in speech. If you would *say* 'it's' or 'won't', *write* it's' or 'won't', not 'it is' or 'will not'. Otherwise you're liable to get it wrong at the microphone.

A NOTE ON NEWS BULLETINS

By our criteria a news bulletin is less radiogenic than a talk. To make its effect, news depends far more on its subject-matter than on its written style or the personality of the man who puts it over. Indeed, the cult of personality among radio newsreaders has always been actively discouraged, at least in the BBC and organisations in the BBC tradition. And yet (see Chapter 5) total impersonality is neither possible nor desirable. Further, radio news bulletins, like talks, are meant to be *heard*. Many, if not most, of the rules for drafting a scripted piece apply with equal force to composing a news bulletin.

3: THE SCRIPTED PIECE: THOUGHTS FOR PRODUCERS

The important thing to remember about talks production is that it can do more harm than good. Speakers aren't actors: they are required merely to be themselves. But this is much more difficult than it seems, given the tensions of the studio situation. Tough, unsympathetic direction will at best turn a speaker into an uneasy imitation of someone else, usually the producer. The effectiveness of a scripted piece, we've agreed, largely depends on the personality of the broadcaster. The producer's function is to make that personality flower, to show itself to advantage.

A talks producer must be an ideas man because he has to father so many separate broadcasts on disconnected subjects. Newly-appointed producers can live on their mental capital for a while, but it soon becomes exhausted unless they make fresh deposits by way of reading and personal contacts. This truth is worth the notice of producers in developing countries, who are commonly overworked, who find it hard to get hold of new books, and who are in any case so much better-informed than most of their listeners that they easily become complacent.

Ideas for talks, or complete scripts, often arrive out of the blue. However attractive they may seem, no producer should accept them until he's certain their authors can speak decently at the microphone. You may think this point too obvious to mention. In fact scores of perfectly dreadful speakers have been put on the air on the strength of an idea, a reputation, or a submitted script. Formal voice-tests may no longer be in fashion, but you can generally lure a would-be contributor to the telephone, when a few minutes' chat will tell you accurately enough what he sounds like.

Anyone you *invite* to broadcast should be briefed. Even a radio trusty will welcome some notion of what you want him to

31

do; with other contributors you should discuss the project in detail. After all, you are the impresario; only you can tell what kind of audience is being aimed at and if the talk relates to the rest of a series or other items in a magazine. In the case of important or difficult broadcasts a *written* briefing usually pays off, if only by obviating argument at a later stage.

Scripts, whether submitted or commissioned, invariably need working on. But there's a right way and a wrong way of revising. Some producers go berserk: they do wholesale rewriting jobs, making the style more colloquial, cutting great swathes through the verbiage, simplifying the argument, altering the sequence of ideas. The outcome may be a model piece of Writing for the Ear, but in the process the contributor may be affronted and his confidence shaken. Time permitting, it's best to sacrifice your ego and work *with* the speaker. As you go through the script you'll notice various feeble, turgid or otherwise unsatisfactory expressions. When you come across one of these don't instantly propose (still less write in) something better: ask your contributor how he would put the point in conversation. The odds are he'll come up with a version that's better still, a phrase you'd never have thought of. And no wonder: he or she is the one who's talking from experience. I once produced a charming, funny, rather naïve talk by a girl about her holiday in Jordan. In draft the third paragraph began: 'With the well-known hospitality of the Arabs I was at once presented with food.' ' "Food" doesn't sound very exciting,' I said; 'what exactly was it?' After a moment's thought she said 'A huge dish of grey, messy, strange-flavoured spludge.' Not a very flattering phrase, but so utterly right for her to say; and as I wrote it into the script I realised I could never have invented anything half so descriptive.

The formula works even when you are faced with unintelligible passages in scripts by experts. Head-on collisions with these people get you nowhere: the more you insist that a given passage must be simplified the more the expert will insist that it can't be done. Try saying as innocently as possible, 'I don't follow this paragraph: what does it *mean*?' Ten to one he'll explain in words of one syllable. All that remains is for you to ask, with appropriate deference, 'Couldn't you say that in the script?'

By the time you and your speaker walk into the studio a lot of your work should be over. But the whole enterprise can still be wrecked – for instance, on the rock of over-conscientiousness. Some producers after hearing the first run-through are visited by far too many happy inspirations for further revisions of the text. That these may be good in themselves is not the point. No speaker can be relaxed and natural, or give free expression to his personality, if his typescript is festooned with last-minute cuts and alterations. In such circumstances he's not concerned with communication: it's all he can do to get the words out. So, when improving a script in the studio, a producer should always know where to stop. Like politics, radio production is the Art of the Possible.

Another mistake is to treat amateur speakers as though they'd been to Drama School, telling them (sometimes even showing them) how every phrase should be uttered and tossing references around to stress, inflection, pitch and pace. This is a guaranteed method of making non-professionals self-conscious. If you find your contributor emphasising wrong words or falling into a sing-song style or a metronomic tempo, you should cure him not by detailed direction but, as previously suggested, by getting him to concentrate on the *meaning* of what he's saying. If he does that the rest will follow. Apart from every other consideration, the alternative method takes far too long: its *reductio ad absurdum* is recording a scripted piece sentence by sentence. This has actually been done in the BBC – a bizarre example of totally synthetic communication.

If you can persuade your speaker to note (and perhaps mark on his script) an occasional phrase that needs stressing, an occasional point where he should pause or introduce a change of tone, you've gone as far as you should by way of detailed direction. On the other hand, don't rush to the opposite extreme and comment on nothing but the timing. Speakers remain uneasy until you assure them they're coming across properly.

Your main, positive, task now is to get the atmosphere right, to *help* your contributor to feel he's not just reading words in a void. How to do this is every producer's secret: there are no general rules but lots of private recipes. One producer I know gets his speakers to talk as quietly as possible, on the theory that this reduces tension. I myself have found that speakers who can't communicate with an invisible audience often improve surprisingly when given a visible one. Many's the speaker I've sat opposite, with an expression of intense interest on my face, during transmission or recording.

I contend, though, that for the most part a producer's place is in the cubicle, not in the studio. As a young man I often directed Outside Broadcasts from variety theatres, which still maintained a precarious existence. On Monday evenings I would attend the first house and choose the acts. During this process I frequently shut my eyes and kept them shut until some potential broadcaster had made his exit, for I'd already learned, the hard way, that troupers whose total performance was marvellous could 'die the death' on radio, where they were unseen. It amazes me that so many talks producers fail to take comparable precautions. In all kinds of ways a radio producer's programme judgement is vitiated by vision. If you are to function properly you *must* at some stage of the rehearsal go into the cubicle, draw down the blind, and simply *listen*.

And I mean listen, not proof-read. I hate to see a producer sitting at the console with his eyes glued to the script, ticking off the minutes and alert to spot the slightest departure from the approved text – as though it mattered – but oblivious of the general effect. When something that looks all right on the page *sounds* terrible – unintentionally comic, or obscene – it doesn't register in the least. *The listener has no script.* Why not put yours down now and again?

FLUFFS

Some talks producers get very worked up about 'fluffs', or slips of the tongue. This is quite unnecessary. The fact is that an interested listener hardly notices them, any more than he notices the weird syntax of an ad libbed interview. However, we can't avoid the subject of fluffs altogether. What should we do about them?

The really crazy course (a popular one, alas) is to say to your speaker just before the recording starts, 'If you fluff, make a pause, go back to the beginning of the sentence, and start again'. It may be that he hadn't even thought of fluffing; so you're 'putting ideas into his head' and making it practically certain

34

that he *will* fluff. What's more, the artificial nature of what you suggest will destroy the atmosphere of communication which you yourself have tried to build up. Speakers should be concerned with conveying thoughts, not with reading sentences.

For my part I never mention fluffs unless the speaker raises the matter. Should he specifically ask what to do if he makes a slip, I say 'Do as you would in conversation: correct yourself and carry on'. This advice removes the anxiety from his mind; so fluffs, generally a result of tension, are less likely to occur. If, nevertheless, he makes a really disastrous slip I get him to re-record the relevant passage *when he's finished the talk*.

This technique, I admit, makes for more complicated tape-editing (not merely finding the fluff and cutting it out but also finding the corrected phrase, cutting that out and sticking it into the gap). However, good communication is more important than the convenience of tape-editors – unless, of course, time presses, in which case a lot of excellent principles go by the board.

POSTSCRIPT ON NON-REHEARSING PRODUCERS

In the Current Affairs area of the BBC there are producers who claim that theirs is a special case; that because of the speed of the operation they have no time to rehearse their speakers. What they really mean is that they don't know *how* to rehearse them. And this is not surprising, because, coming as they mostly do from newspaper journalism, the producers concerned don't understand what a rehearsal is *for*: they regard it as a meaningless formality, which in their case it is. So they fall back on the expedient of recording the first read-through and then (since even they can sense that this is seldom satisfactory) doing numerous retakes and patching them in to the tape. By the end, of course, the process takes as much time as a proper rehearsal-and-recording. So as producers they have gained nothing and lost a great deal (I mean in terms of programme quality: the waste of tape, and of editing facilities, is a separate issue). We are dealing here with nice, able people. It's a shame that because of the BBC's haphazard methods of selecting staff for training so many Current Affairs producers have escaped instruction in the very elements of their new craft. Perhaps this book will help to fill the gap.

Not everyone who has something to say can write a script. But we are all capable of answering questions. Hence the radio interview. Let's deal first with the commonest kind, lasting only a minute or two and meant for a magazine or a news bulletin. Here the interviewer plays his basic role of getting the other person to say what will interest the audience, and to say it in proper sequence and for the right length of time. The interviewer is a Sherpa, a Prisoner's Friend, a radio midwife.

Successful interviews aren't happy accidents: they require preparation. One must first be clear about the object of the exercise. Anybody commissioning an interview should explain *why* A is to be questioned, what the story is, where exactly A fits into it. A precise directive gives an interviewer something to aim at. Vague, waffly directives lead to vague, waffly interviews, which then have to be painfully edited into shape. Once briefed an interviewer starts to assemble background information. Often there's little time for this stage, but one should try not to omit it altogether. Handouts, press cuttings, reference books, phone calls – all these can be useful. Your aim should be to arrive for the interview knowing enough about the interviewee to make yourself agreeable, and enough about the subject to frame intelligent questions of your own instead of depending entirely on the suggestions of the editor.

Above all, you should arrive with a *plan* in your head. You will probably have to modify it either during the preliminary chat-up or – more unnervingly – while the interview is in progress. But an amended plan is far better than no plan at all. Beginners, by the way, almost always overestimate the number of points they can deal with in a given time. Preparing an interview gets easier with practice. Interviewers, like barristers, learn to absorb the main facts of a human story at speed, retain them as long as may be necessary, and then put them out of their minds. (What this does to their finer feelings is another matter. As Antony Jay wrote in *The Listener*, 'Every interviewer must from time to time wonder whether he is not a licensed confidence trickster. Through sympathy and insight he extracts quotes whose value and impact are far more obvious to him than to the interviewee who supplies them, and the personal relationship which gained them was possibly taken by his subject as the start of a friendship, whereas for him it was simply the day's assignment.')

THE CHAT-UP

The most important part of an interview is the few minutes before it starts. If your victim takes to you he's likely to respond more freely. So the impression you make is crucial. You may be tired; the subject of the interview may bore you: if so, don't let on. Make it appear that this is one of the richer moments in your life, that you're keen to know what the interviewee has to say. Concentrate on setting him at his ease. An interviewer has to command a certain acting ability – though if you have to call on it very often you're probably not in the right job: one point on which the whole tribe of interviewers agree is that to succeed you must feel a *genuine* interest in other people, if only as specimens.

The chat-up is important in another way: it tells you something about the interviewee's style of speech. Once you get his speech rhythms into your head you'll be less likely to talk across him, or alternatively to come in so late with your questions that the broadcast will be full of holes.

Should an interviewer tell? How far, if at all, do you disclose the line you propose to take later on? Sometimes you have no option. Your speaker may demand notice of your questions, and if he's a notable in a developing country he'll be in a position to insist. If he does, all you can expect is a parody of an interview, with yourself cast in the role of stooge, feed or straight man. If, on the other hand, you keep your intentions entirely to yourself you're liable to elicit some very confused replies, especially if your victim is a nervous type, wondering all the time what's going to hit him next. Some of his answers may even take you by surprise and throw you off balance.

All things considered, it's usually best to 'define the area of discourse', i.e. to agree with your victim on the main points to be covered, and to stop there. If you go over the ground in detail the interview will be short on spontaneity. The answers will be beautifully compact, but the significant details and revealing asides that come pouring out the first time won't be repeated: what you'll get is a précis. You may even have to remind your victim of what he said beforehand, and as an interviewer you can't sink much lower than that.

It's often useful to devote some of the chat-up to points *connected with but not the same as* the subject of the interview. This gets a speaker's mind working in the right direction without plunging him into a rehearsal situation.

RUNNING THE INTERVIEW

The confidence you have created is a frail flower, which mustn't wither during the interview itself. For some reason *spoken* expressions of interest – 'Really!', 'How extraordinary!', 'H'm, h'm' and the like – sound silly in the mouth of an interviewer. It's all the more important to signal encouragment in the only way left: by your expression. Keep looking at the interviewee, smile at appropriate moments, nod approval. This may seem childish, but it works. Don't keep darting anxious glances at your recording machine: once you've set the controls after a voice-test, forget about them and trust in Providence. An unwatched machine may sometimes let you down; an interviewee who senses that you're only half-listening will *always* let you down.

It should now be obvious that notes are a nuisance. They come between you and the speaker. In a short interview you shouldn't need them at all, but if you must have them let them be confined to single-word headings (on a stiff card). There's something to be said for scripting the first question (if you're incapable of learning it off by heart) because a smooth, decisive opening is always an advantage. But to write out *all* your questions is to court trouble. The results are invariably dreadful: interviews proceed in a series of jerks; no question seems to arise naturally out of the previous answer; and rapport between interviewer and interviewee is non-existent.

Here are some practical suggestions. First, since written 'cue material' (information for the presenter of the programme) can get mislaid, it's as well, before starting a location interview, to *record* some identifying words – your name, the speaker's name, the date and so on. Leave a gap on the tape between all this and the interview proper: otherwise some operator in a hurry may transmit the whole thing.

Sometimes you have to act as your own presenter and introduce the speaker at your side. This can be embarrassing for him if you go on too long. Make your introduction *relevant*: tell the listeners all they need to know but don't go into superfluous detail or arouse expectations that won't be realised. And make it *attractive*: give your listeners some cause to stay with you. Finally, phrase your introduction as *freshly* as you can: we're all sick of 'With me in the studio [or on the tarmac, or wherever] is Mr So-and-so, who has done such-and-such. Tell me, Mr So-and-so . . .'.

Listeners don't like being kept in the dark. If you know that your interview will take place in an odd acoustic or to the accompaniment of unrecognisable noises-off, explain in your introduction (or cue material) where you are. By the same token, if some unexpected noise starts up during the interview itself don't ignore it. Account for it and press on (or, when the noise has stopped, go back to the beginning of that particular section and start again).

When you begin the interview itself *get to the point* as quickly as you can. Don't embark on some wide-ranging survey from which your theme ultimately emerges if you're lucky. It's a waste of precious time. So are questions whose only object is to elicit information that ought to have gone into the cue material. Having got to the point, stick to it. *One subject is enough for one interview* (of the sort we are considering).

Of course, all interviews are unpredictable. Once the recording is under way a speaker, to his own amazement as well as yours, may go off on some totally unexpected tack. What should be done? Usually one tries to steer him back on course; but if the new line of thought is interesting you might be wise to let him pursue it, working towards a decent conclusion as best you can. Everything depends on circumstances and your own mental agility.

Remember that, as in the case of talks, most listeners – English listeners anyway – are bored by generalities and prefer what is specific and concrete. Let's say you're interviewing a recently-elected Council Chairman. 'What are your hopes for an improvement in the quality of life in our city?' is a less arresting question than 'Can you do anything to get the buses running on time?' If your *victim* is a generaliser try to pin him down. Few questions are more useful than 'For instance?'

Also as in the case of talks, the personal element should never be lost sight of. Don't trivialise important issues, but remember that even in straight news interviews if you don't exploit your speaker's personality to some extent you are probably throwing away your trump card. Try to establish what makes your Council Chairman *want* to be a Council Chairman: what impels him to endure the tedium of committees and bad dinners. If that would take too long, why not ask how he plans to combine his official duties with the running of his business?

Make your questions as succinct as possible. For instance, a full statement of alternatives is usually quite unnecessary. Let's assume you're interviewing a country postman:

Do you find it tiring to walk nearly sixteen miles a day, or have you got used to it by now?

is too long.

Do you still find it tiring to walk nearly sixteen miles a day?

says it all.

Above all, make your questions *clear*. Your interviewee is surely entitled to understand what you're trying to ask him. A question beginning 'Can you tell me something about . . .?' deserves the answer 'What do you want to know?' I only wish more people would give it. Double-barrelled questions should always be avoided:

From the financial point of view companies must to a certain extent be ruthless, mustn't they, and when the orders drop off get rid of workers? Do you find it hard to be ruthless?

Which question is the 'ruthless' interviewee expected to answer?

If your speaker is a man of few words, who replies to every

40

question with a straight Yes or No, remember the *Who, What, Why, Where, When and How* rule, invented by some unknown genius. If you start a question with any of these words it's impossible for the interviewee to answer Yes or No. Especially useful is 'Why?'. It demands *reasons,* to provide which an interviewee is almost forced to spread himself a bit. Another way to get extended answers is very simple: *pause and look expectant.*

If, as seldom happens, your speaker talks too much, don't rudely interrupt him. Look him firmly in the eye, and when you sense that he's about to draw breath 'arrest his attention with a gesture' – an uplifted index finger does very nicely. This will stop in his tracks everyone but the most hardened bore and enable you to slip in your next question with some aplomb.

Sometimes an interviewee deploys technical terms or makes obscure references. If your audience is a general one your best course is to ask for an explanation right away.

If you happen to know a lot about the subject don't make a parade of your expertise. A chat between equals is not an interview.

Never repeat what you've just been told – thus:

Answer: . . . so I went by train.
Question: So you went by train. In that case . . .

This kind of verbal tic is maddening to listeners.

We've mentioned the need to start well. Try also to *finish strongly.* End with a bang, not a whimper. And don't fade away with a lot of insincere thanks: unless the circumstances are exceptional the time to thank your speaker is when you've switched the recorder off. In any case don't end by saying 'Mr John Smith, thank you'. What everyone says in real life is 'Thank you, Mr Smith'. The other way of putting it was started by BBC TV and has spread like a rash over the mass communications of the English-speaking world. For my money, the ideal is achieved when an interviewee says something so conclusive that further questions are superfluous, and when the interviewer has the sense to realise this and shut up.

ALL AT ONCE

An experienced interviewer at work is doing a great many different things. He is genuinely listening to the other person. He is continuously helping that other person – by showing interest and taking up his points quickly. The interviewer with another part of his mind is cold-bloodedly acting as his own critic, assessing the development of the interview as a piece of radio, controlling and shaping it, adjusting his attitude to the interviewee in order to get the best broadcasting results. He's keeping an eye on the stop-watch: he's preparing his next question. And he's doing all these things at once. Interviewing is quite a craft.

EXAMPLES

Here are examples of two very short interviews on the same topic: the treatment a coloured immigrant received in an English city. First, how *not* to do an interview.

Link-man's introduction, including name of visitor, description ('a student'), country of origin, name of interviewer.

Question 1: When did you arrive in . . . ?
Answer: Two years ago.
Question 2: And where are you studying?
Answer: The Polytechnic.
Question 3: What subjects are you studying?
Answer: Chemistry, Maths and Physics.
Question 4: Why did you choose these particular subjects?
Answer: . . .
Question 5: Where are you staying, in a hostel or in digs?
Answer: In digs. A couple of miles from College. Quite handy really.
Question 6: Do you like your digs?
Answer: They're not bad.
Question 7: Did you have much difficulty in getting accommodation?
Answer: Yes, I suppose I did in some ways. But then it's not very easy for students to find digs.
Question 8: Um – er . . . Would you say there was much racial prejudice in this city? I mean, have you run up against it yourself?
Answer: Er – no, not particularly.
Question 9: (*Pause*) How much longer do you expect to stay at the Polytechnic?
Answer: Two years, if I get through my examinations.
Question 10: Oh, I'm sure you will. . . . Well, the best of luck. Thank you.
Answer: (*with obvious relief*) Thank you.

42

'A lousy interview,' the producer will say. And he'll be right. The interviewee may get the blame. But of course he never had a chance. Let's analyse the interviewer's technique and pin-point his crasser mistakes.

Questions one to three are unnecessary. The programme has a link-man: why not let *him* include these facts in his introduction? It can be done neatly: it will take less time, and the interviewer will be able to start on an *interesting* point. Question four is open to more than one objection. If the interviewee replies that he is studying some highly unusual combination of subjects the next question will have to be 'Why?'. We are then off on an explanation of the special needs of this student or of his developing country. It might be fascinating but it has nothing to do with the theme of the interview. On the other hand, if the student names, as in fact he does, three very ordinary subjects, the answer is not only irrelevant but dull. The interviewer now compounds his errors by chasing what is obviously a red herring through a further question and answer.

Eventually, in question five, we come to the point. But this and the next two questions are not very subtle. In answer seven the student is clearly reluctant to allege that he has been discriminated against on racial grounds. The interviewer could have followed up with 'Are you saying that it was no harder for you to find digs than for an English student?' This isn't a loaded question; but it would have compelled the student to say what he really thought, and thus have carried the argument a stage further. Instead, the interviewer pauses and tries a different approach.

Question eight can be faulted on several grounds. It is double-barrelled: it demands both an opinion on a general point and an account of personal experience: no wonder the student takes refuge in a meaningless reply. Again, the real subject of the interview, racial prejudice, is introduced much too suddenly – hurled at the respondent like a hand-grenade. His first instinct is to duck – in this case to duck the question. By this time our interviewer realises the game is up. So he falls back in disorder on an earlier line of questioning (which was itself beside the point) and somehow brings the interview to a lame conclusion.

The interview *could* have gone like this:

Link-man's introduction, including name of visitor, description ('a student at the Polytechnic, who came here two years ago from . . .'), name of interviewer.

Question 1:	Where are you living now?
Answer:	In digs. A couple of miles from College. Quite handy really.
Question 2:	Where exactly?
Answer	. . .
Question 3:	That seems an unusual area. Where else have you had digs?
Answer:	. . .
Question 4:	Quite a lot of places. Why have you moved about so much?
Answer:	Because I've never been satisfied with the digs I was in.
Question 5:	But why did you go to such places?
Answer:	They were the only ones I could get.
Question 6:	Does that mean it's harder for you to find decent digs than it would have been for a white student?
Answer:	Yes, I'm afraid it does.
Question 7:	Can you prove it?
Answer:	Yes, I can (*gives instances*).
Question 8:	Well that sounds like racial discrimination all right But has *everyone* you've met been prejudiced?
Answer:	Oh, no (*gives instances to the contrary*).
Question 9:	Well, I'm glad to hear *that*.

This time the questions are strictly relevant: in the few moments available the interviewer goes logically from point to point, constantly digging a little deeper, until a conclusion is arrived at in the answers to questions six and seven. There is still time to bring that picture into balance by the answer to question eight. Plainly this is interviewing of a much higher standard.

INTERVIEWING AT LENGTH

In *longer interviews* the questioner is much more than a guide, or radio midwife: he, so to speak, examines the other party on behalf of the public. He can still be basically sympathetic but his scalpel cuts to the bone. I remember producing a long interview – before editing, it lasted three hours – in which the novelist Richard Hughes was slowly grilled over a fire of questions about his personal motivations and working methods. The tone of the interviewer (the critic Walter Allen) was one of admiration. But

I noticed that as Allen settled to his task, and as each man took the measure of the other, enquiries on delicate matters began to be pressed with a freedom that would have seemed mere loutishness in a briefer broadcast. On this occasion Walter Allen, who has read everything and remembers everything, didn't use a single note. But such a feat would be beyond most of us. Much as one deplores the interviewer who depends on notes for a three-minute recording, it must be admitted that you normally can't do without them when you're working in depth. Even so, they should be headings and nothing more.

The *tough interview* is the final, though not of necessity the most admirable, stage of the interviewing technique. In many countries the idea of submitting a person in authority to a genuine interrogation on the air is inconceivable. Until fairly recently it would have been inconceivable in Britain. And even today the British broadcaster who practises this activity has to be careful how he goes. It's not merely that ethics and the law of libel enter into the question. It is now apparent that tough interviewing can be counter-productive, swinging the sympathies of the audience irresistibly behind the victim. Why this should be so is not for me to explain, though I could make a few guesses. But there's no doubt that once a certain limit has been passed the transfer of sympathy is immediate.

Not long ago it became briefly fashionable to harass and insult such interviewees as couldn't retaliate, on the theory that once shaken out of their self-possession they would give more revealing answers. But it was soon found that the 'needling technique' needled the audience even more than the victim. In a probing interview it's sometimes necessary to ask very plain and direct questions, and if the interviewee is evasive to repeat them until either he comes clean or it's apparent to all that he's refusing to give a straight answer. But, unless you think you have a future as the man the listeners love to hate, the more courteously your questions are put the better. Manifest hostility to the interviewee merely weakens their effect. It's quite enough to ask him for his comments on some damaging charge: there's no need to imply that you agree with it. A searching question neutrally phrased can be devastating.

Interviewing for Documentary belongs to a different world. We'll deal with it in a later chapter.

5: PRESENTATION AND DISC-JOCKEYING

I

A radio announcer is a cross between a microphone speaker and an actor. He is also – bad luck – an official. When I was new to the announcing trade a senior executive took me aside and said 'Do be careful when you go to the mike. Remember you are The Voice of the BBC.' Such spontaneity as I possessed vanished instantly, and took a long time to re-emerge.

The impersonality of BBC announcing was doomed to die. It survived longest in the Third Programme, whose presentation seemed designed to repel rather than attract. All those priggish phrases (Our speaker tonight . . . Listeners may recall . . . In 1743 he was appointed *Kapellmeister* and court organist . . . Dr Smith, Fellow of St Sepulchre's College, Oxbridge, offers a few observations . . .) delivered in those prissy accents – who could believe they preluded an entertaining programme?

BBC announcing, in the domestic services particularly, is now very different. But that senior executive's attitude can still be paralleled within the hierarchies of the newer broadcasting systems. As a result what one so often hears from African stations is mechanical, uptight, *boring* announcements, totally at variance with the natural exuberance and charm of those who deliver them. It's true that the style in which programmes are presented does affect a network's image, so that staff announcers should feel a sense of responsibility to their organisation. But they can combine it with the ability to sound like human beings. If they sound like automata they fail to communicate and therefore fail as announcers.

CONTINUITY

When I was young I used to announce in lots of different studios, sharing the tension of performers who were 'going out live'.

Nine programmes out of ten are now pre-recorded, and my successors are mostly marooned in air-conditioned continuity suites. Under such conditions it's easy to lose touch with broadcasting reality. What a staff announcer must bear in mind is that the bits of equipment around him aren't elements in some private game, that they result in a flood of music and speech which real people are listening to moment by moment, that when he fades up his microphone he's not just 'making an announcement' but actually talking to somebody. If he forgets he soon becomes yet another broadcasting zombie, uttering 'polite meaningless words' like a man talking in his sleep – with occasional ghastly over-compensation.

In the newer countries of the Commonwealth the continuity announcers of the BBC's External Services are often taken as models. This is unfortunate, for good though these announcers are they labour under peculiar difficulties. They do much of their work in the dead of night, when few communicators are at their liveliest. They are remote from their audiences, who are in any event too scattered and diffused to be imaginatively visualised. And in order to cut through the fading and distortion inseparable from short-wave transmission they are compelled to mouth their words in a quite unnatural way. It's a pity that African and Indian broadcasters so seldom hear the BBC's domestic presentation.

A continuity announcer, as his name suggests, is responsible not merely for introducing specific programmes but for holding together a band of broadcasting, giving it the proper feel, and, in stations of a mixed character, for modulating agreeably from one style of radio to the next. This he can't do if he doesn't *listen* to what is being transmitted. 'Easier said than done,' he may retort, pointing to all the mechanical operations he has to carry out. But I fancy that many continuity announcers could listen more attentively if they tried. So often (perhaps because they've been doing the job too long) they mentally switch off between opening and closing announcements. Hence all those tiresome lapses in communication – inappropriate trails, fill-ups that clash horribly with the preceding programme, and closing announcements that come in too quickly, ruining some drama producer's carefully-contrived climax.

Some smaller stations are now making bad continuity work

a virtual certainty by going in for 'automated presentation'. So-called continuity announcements, together with trails and commercials, are pre-recorded on cassettes: these are slipped into a machine and started off (by an engineer) as required. The results are what might be expected.

SELLING

The chief function of any announcer is to *sell* the programme – to a particular audience in each case. So you need a flexible approach. If you sound much the same whatever programme you're introducing you are probably not very good at your job. You need to be equally flexible in your choice of *information* for opening announcements. Facts in themselves – whether about speakers or musicians or the music to be played or any other programme ingredient – are often not very helpful. Select the information that will pull in an audience and help it to enjoy what it hears.

It's always a temptation to be personal in the wrong way, to draw attention to yourself rather than the programme material. Sometimes it's legitimate to do this, e.g. when you're putting on Pop discs, but with programmes in general your métier is to provide a frame for someone else's picture. It's also a temptation to overdo the ad libbing (on those rare occasions when you get a chance to ad lib). I remember listening to a certain announcer presenting records. In two hours he began *twenty-seven* sentences with the words 'And now' (in fact he must have said it oftener, because I only started counting when the phrase had got on my nerves). He was good in every other way. Had he condescended to script just a few of his impromptus he would have been first-rate.

READY-MADE ANNOUNCEMENTS

Like Bob Hope and the late President Kennedy, every announcer has an army of ghost-writers – the producers who supply him with ready-made announcements. Often these are highly predictable ('We present . . .'). They may be full of irrelevant information or unattractive in other ways. What does the announcer do? It's sometimes possible to improve the wording.

48

You can go carefully through the texts and alter them just enough to make them suit both the programme and your own style. This is a laborious process; you have to spend a lot of time telephoning producers to explain what you have in mind and to ensure that your amendments haven't destroyed some point of substance. But it often works. In extreme cases you can compose a completely new form of words and put that forward as an alternative. Most producers, regrettably, know nothing about presentation and regard the construction of announcements as a time-wasting chore (some of them leave the whole thing to their secretaries). So they're usually happy to accept suggestions. One or two might even thank you for taking so much trouble.

In the end, of course, whatever you may think of the announcement (or the programme), you deliver your lines with all the conviction you can summon up. This is when you often have to start acting.

NEWS

When you read the News your script has always been put together by another. And you're not allowed to change a word of it. Furthermore, now that news bulletins are interlarded with recorded reports and interviews, you are taking part in a production. Nevertheless, *the chief virtues of a newscaster are those of a man giving a talk, not an actor.*

The newscaster's job is to study the text, to absorb the meaning (sometimes at speed) and, while sticking to the words provided, and without letting his own feelings obtrude, to *tell* the audience what the news is. If he goes to the microphone with a fully-marked script, intent on giving a splendid reading, he usually gives a disastrous one, communicating next to nothing. Reading out the phrases rather than concentrating on the sense is also, incidentally, the surest way to develop irritating vocal mannerisms, such as the piledriving thump which some newscasters bestow on the last word of every sentence, whether important or unimportant. For instance, 'The pound had a better day, and share prices went up in the *City*.' Where else?

It's a good idea to practise looking *ahead*. If you can do this easily you'll be saved from many a gaffe when you have to cope with paragraphs you haven't had time to examine before going on the air.

Many newscasters sound monotonous simply because their speech-rhythms are too regular. As we know, in ordinary speech words come in bursts, with longer or shorter pauses between them. And the tempo of ordinary speech is constantly varying, some phrases, in accordance with the meaning, being dwelt on much longer than others. If you are a metronomic reader try leaving one hand free and using it to make gestures as you go along. You'll be surprised how much this helps you to break up the material into speakable phrases.

ADDITIONAL THOUGHTS

When you're at the microphone you should be reasonably relaxed – if you're too tense a few deep breaths will make you less so.

In a continuity or news studio sit comfortably, but let your posture be upright. Don't look down, with your shoulders hunched, your chin dropping, your throat muscles constricted and your chest bent inwards. And don't lean forward. Sit up, keep your chin up, and direct your speech to an imaginary listener about four feet away from you. This is how to get the necessary degree of projection.

You will often have to broadcast while wearing headphones (cf. the section on compering magazines). Beware, in these circumstances, of starting to listen to your own beautiful voice. (On headphones it always sounds richer and grander than it does in real life. It's like singing in the bath.)

So far I've carefully kept off the subject of accent, which at one time obsessed announcers and those who selected them. In Britain it's coming to matter less and less. But one observes that in Africa and Asia continuous efforts are still being made by relays of advisers and expatriate instructors to get domestic announcers to talk like Englishmen. This may be admirable from the standpoint of Commonwealth unity but it has singularly little to do with effective communication by radio. Young Africans and Asians may succeed in modifying their vowels *pro tem*, but the result of this emphasis on phonetics is that they become damnably self-conscious and the life drains out of their announcing. If *domestic* announcers in the New Commonwealth sound like educated citizens *of their own countries* that's surely good enough. Isn't there more than a whiff of outmoded cultural imperialism about the whole accent-bending enterprise?

II

DISC-JOCKEYING

Freelance presenters of record programmes have been around for a long time – Christopher Stone was at it well before the War. Some student of mass communication really ought to give us a history of their activities. It would include chapters on Stone and his kinsman Sir Compton Mackenzie; on the stars of Radio Normandy, Radio Luxembourg, Radio Lyons, Poste Parisien and the other stations that beamed sponsored programmes to Britain in the days of Sir John Reith; on the influence of the American Forces Network and the importation of the term disc-jockey; on the post-war renaissance of Radio Luxembourg, and on the new wave of DJs associated with the pirate stations – as well as on more recent figures and events.

In all this time the DJ's role has remained very much the same. It is to do the presenting in a more individual style than staff announcers normally go in for. At the Pop end of the spectrum, strongly influenced by U.S. radio, the emphasis on individuality has increased so much that the Pop DJ in fact provides a show, in which the exploitation of his personality is as important an ingredient as the music.

To succeed as a DJ the first essential is that a large, and for the most part youthful, audience should be able to identify with you. Why one broadcaster can bring this off when another, equally bright, equally articulate, and possibly much nicer, can't, is a mystery. But without this x factor, this charisma, no DJ can make the grade. Closely allied to it is a kind of sincerity. Very few professional DJs, however long they may have been in the business, deliberately talk down to their audiences: while they're on the air, at least, all they live for is whatever brand of popular music they happen to be presenting. (Veteran status, by the way, is not a disqualification. Some older DJs find it prudent to migrate from Radio One to Radio Two, but others, perpetual adolescents in spirit, continue to shake their dyed locks all over the BBC's Pop channel. Personally I can't help wondering if the 'kids' are quite as unconscious of age gaps as programme-planners imagine. Capital Radio's speedy climb to favour would suggest otherwise.)

A DJ needs a smooth tongue. You must be able to put words together unhesitatingly, with minimal help from notes, while executing a complicated series of manoeuvres with discs, cassettes, cartridges and faders (manual dexterity is another essential qualification). A pint-size vocabulary is no use: it's a sign of weakness to use the same expressions over and over again. I'm not referring to catchwords, be it understood: employed with discretion these can be valuable trademarks. What one should avoid is commonplace phrases ('spinning a disc') and epithets ('great', 'super', 'fab', 'fantastic') which if repeated too often start to irritate the listener as acutely as my announcer's 'And now'. A DJ I know well tells me that when he was a young man, aware of gaps in his vocabulary, he set himself to learn five new words a day. It seems to have paid off (just as it paid an OB commentator to do much the same – see Chapter 13). If you suspect that you may be getting into verbal ruts, why not have one of your sessions taped, and then listen to it in cold blood?

At one time British DJs went in for American accents. Perhaps they wanted to be in tune with their material, for the most notable popular music then came from the U.S. That situation has changed, and, coincidentally or not, the bogus mid-Atlantic style is on its way out – in Britain, that is. Unfortunately the habit has spread: I heard only recently a DJ from proud Northern Nigeria doing his best to sound as though he'd been brought up in Harlem. As listeners become more sophisticated this sort of make-believe cuts less and less ice. In Britain today what gives a DJ his appeal is chiefly liveliness of mind and manner; an ability to be himself, his own man. Accents range from cut-glass through 'classless' to near-gorblimey, and no one cares in the least. I find this an exhilarating development.

KNOWING YOUR STUFF

Unlike most of the celebrities who are occasionally brought

into a studio to introduce records, professional DJs have a wide and exact knowledge of popular music. Their presentation is firmly based on it. I say 'wide' because a knowledge of certain areas only is not enough. Some acquaintance with classical music, even, is not to be sniffed at, for audiences are tending to be more catholic in their tastes: it's quite common to hear of listeners who switch on Radio Three as readily as Radio One, Capital, Medway or Hallam. A few DJs have noted the trend, and make, so to put it, classical allusions now and again.

Let me stress, in passing, that to be well-informed about music, especially Pop, will not *in itself* qualify anyone to become a disc-jockey. This should be obvious, but in my experience many people, unhappily for them, don't get the point.

Another caution. To be keen on Pop is essential; to be starry-eyed about it, fatal. Pop music is usually exciting, or pretty, or sexy, or attractive in some other way. But it's not only an art-form, it's an industry, an enormously profitable industry, geared to tapping the new affluence of the young, and dominated by capitalists of the type Sir Harold Wilson delighted to honour. Record companies employ pluggers, alias promotion men, and these chaps are not over-fastidious in their methods. Once you make your mark as a disc-jockey they'll be after you. Do remember that if you sup with them you will need a long spoon.

But that, as they say, is to anticipate. In Britain the difficulty has always been to get your foot in the door, to move from the underworld of campus radio, hospital radio, ship's radio, mobile discos and the like into broadcasting proper. Most of the BBC's top DJs owe their first break to the British Forces Broadcasting Service, Radio Luxembourg, or the pirate stations. But the pirates have been chased off the sea, the role of the BFBS is changing, and Luxembourg is not quite what it was. Fortunately, local radio, BBC and Independent, looks as though it might close the gap. Stimulated by a lack of funds, lots of station managers are positively welcoming talented unknowns.

IT'S A MAN'S WORLD

Getting on the air becomes more difficult still if you happen to be a woman. The female disc-jockeys we have in Britain are uniformly excellent and most habit-forming, but there are astonishingly few of them. Why this should be I don't understand, especially when the air is full of women newsreaders, not all of whom carry great conviction. Disc-jockeying really does merit the attention of Women's Lib.

VARIETIES OF DISC-JOCKEY

DJs fall into different categories. Those who operate on the BBC's Radio Three and Radio Four are firmly fixed in the Christopher Stone tradition. Then there's the Radio Two school, concentrating on 'light music' – anything from Sigmund Romberg to 'The Cornish Floral Dance'. At the moment Jean Challis (ex-BFBS) is probably the best of the Radio Two DJs. Warm, elegant and well-bred, she could easily be imagined hostessing 'Woman's Hour'. (Light music presenters seem able to slide into the Current Affairs/Hearth and Home department quite easily, witness Jimmy Young's superb live interviewing of Cabinet Ministers and other unlikely characters in between discs.)

Pop DJs can be variously subdivided. You can classify them, for instance, by type of material. Some are Top Twenty merchants: they go by the charts and reflect the run-of-the-mill aspects of the current Pop scene. Others specialise – in Soul and Blues, Folk, Country and Western, Hard Rock, or whatever it may be. A more (indeed totally) subjective distinction is between those few who overdo the personality bit and the Rest. America has given us the kind of DJ who sounds as though he were 'high', who goes maundering on, sometimes, it seems, for minutes at a time, about whatever comes into his head – food, women, the weather – and who plays the next disc when he remembers to. I find this style totally unappealing, but some people must like it.

Another possible division is between DJs who show respect for the music and those who don't. Only a small minority now avoid talking over it – indeed, when so many recordings come to no musical conclusion, offering instead only the repetition of an unresolved phrase on slow fade, some 'talking over' is almost forced on you. What matters is how you do it. The sort of DJ who chats away regardless, knowing that the compressor will automatically lower the volume of the disc, who then lets the music surge up at any old point, and who is capable of fading out or even cutting the disc with no thought whatever for the *shape* of the tune – this fellow is a barbarian. And what of the DJ who lets us hear him singing or whistling the melody during some superior passage of orchestration? He should be shot at dawn, or at least kept off the air for a week or two.

In schizophrenic fashion, a skilled Pop DJ not only plays-in the music, not only maintains his smooth flow of chat, but produces *himself*, as performer, and the show as a whole, while it rolls along. The programme may indeed have an official producer assigned to it, and his advice may be useful before and after the event, but during transmission no direction is possible except in the most general terms. So the movement of the show, and in particular the interweaving of music and speech on which so much depends – the precise way a disc is brought up behind the last words of an introduction, the way the last words of a lyric are answered in the next bit of chat, the way a jingle is worked into the over-all pattern, the way normal introductions are interspersed with back-announcements – all this is a true test of the DJ's quality.

A final point. When I recently asked a vastly experienced DJ, now in charge of various major series for BFBS, what was the most important piece of advice he could give a beginner, he replied 'Sound as though you've been listening.' In other words, whatever the distractions, and there are always plenty of them in studio and cubicle, and however much energy you have to expend on working the equipment, you should keep your mind on the music: it's the only way to make sure that when you open your mouth you don't put your foot in it. Much the same advice as I gave continuity announcers, and none the worse for that.

6: A NOTE ON PHONE-INS

The engineering technique of linking a studio by telephone to the outside world and broadcasting the ensuing conversation has been with us for a long while. But in Britain it's only comparatively recently that the phone-in became a programme category in its own right. Its popularity may not have much significance for art radio but it's important nevertheless, for at least three reasons.

As programme controllers in the BBC and commercial radio are well aware, the extended phone-in on any old subject is the cheapest formula yet devised (apart from playing records and forgetting to pay copyright) for filling unlimited quantities of air-space. On the political level this form of 'participation' opens up new possibilities, giving listeners for the first time some chance of challenging the power of media men and interested parties to impose their view of events on the community at large. Thirdly, for people in temporary difficulty or under permanent stress, isolated individuals in great cities, even the degree of understanding and human contact afforded by a phone-in can be valuable. Late at night and through the early hours phone-ins are 'almost entirely a social service for the lonely, especially the insomniac lonely'.

The producer of a phone-in may know in advance – or think he knows – what subjects the callers intend to talk about. But once the programme is on the air he has no control over them. In the pre-packed world of broadcast entertainment this element of unpredictability is an additional attraction. (American stations use 'the loop', a recording device which provides a delay of five seconds or so between utterance and transmission – long enough to enable someone to do a jamming job when necessary. But in the BBC they prefer to rely on intuition. The producer, aided by a bevy of assistants, chooses twenty or so from the hundreds of would-be questioners who have phoned up beforehand, rings them back, puts them on the air in turn, and hopes for the best.)

The *extended* phone-in programme, during which someone in a studio deals, hour after hour, with questions, suggestions, complaints and appeals, often of a highly personal nature, from a string of unseen and half-identified callers, holds an eerie sort of fascination for most of us. It's like bugging a confessional box or, as Alan Brien puts it, listening to other people's conversations on a permanently crossed line. What the central figure must have, in addition to a quick mind and a readiness to pass judgement on the basis of one-sided information, is interviewing skill; for in virtually every case as soon as he's replied to the first question (or suggestion, or whatever) he starts to ask questions himself, with the double motive of helping the caller and interesting the audience.

There is another, perhaps more intellectually respectable, kind of phone-in. In this the proceedings are conducted by a chairman and the questions answered by an 'expert' – sometimes more than one. When, as often happens, the visitor is a skilled manipulator of the media – perhaps a politician, perhaps the head of a nationalised industry or a big trade union – the chairman must be especially alert to see fair play. Alan Brien, noting in the *Sunday Times* that 'it has taken the British a couple of decades after the Americans to realise that radio has many strengths, beyond the power of TV', remarks that phone-ins 'provide a direct, intimate, practical kind of "access" which . . . camera-dominated studio confrontations can never match. A discussion between public and pundits gains when both are equally invisible.' No doubt, but only when both are equally matched. This kind of phone-in, like the other, is a series of interviews, but in this case it's the questioners who are at a disadvantage. It's they who are the non-professionals, they who need protection from the verbal adroitness, the deadly fluency, of the public man. It's he who is familiar with the microphone, who always sounds louder and clearer than the pople on the line, and who has the technical capability of talking right through them if he wants to. Every chairman should copy George Scott's admirable habit of occasionally asking 'Are you *satisfied* with that answer?' and seeing that necessary supplementaries are duly put. The familiar sequence, question/evasive reply/next question, is a mere parody of 'meaningful dialogue'.

As successful a phone-in as I've heard was recently broadcast by BBC Radio London. An intelligent and friendly doctor was dealing with questions on children's illnesses. The girl in the chair sounded concerned and interested; although she kept the programme moving she took care to give him all the elbow room he wanted so that a real dialogue occurred between the doctor and each anxious mother. Perhaps the phone-in is at its most useful when a genuine expert, unhurried and with no axe to grind, encounters genuine enquirers on a well-defined range of subjects.

7: PRODUCING A DISCUSSION

At one time discussions were prominent features on the radio landscape. There are fewer of them now, but they still occur, in a wide variety of durations. Often they make excellent listening. So it's important to know how to deal with them.

The classic type of discussion consists of several people expressing different opinions and arguing among themselves under the guidance of a chairman who remains neutral. So the first essential is a controversial subject, one that generates its own electricity. Without some element of controversy you'll get nothing more than a multiple interview or multilateral chat, interesting or uninteresting as the case may be. A discussion isn't the same as a debate, and there's no reason to confine yourself to Yes or No questions. As a subject 'The Future of Secondary Education' opens up many more vistas than 'Should Public Schools be abolished?'

One should always try to fix on a *manageable* subject, whose scope bears some relation to the air-time at one's disposal. It's no use trying to deal with race prejudice in six minutes. If the subject is too wide, vital aspects will be overlooked, and there will be no time to *examine* the various assertions made. A good discussion is more than an entertaining shouting-match: it should elucidate the issues.

THE CHAIRMAN

So the chairman, or moderator, has a key role. It's his job to keep the discussion moving, and moving along logical lines. Without being a latter-day Socrates he should at least be able to define *precisely* the areas of agreement and disagreement and to confront speakers with the logical consequences of their own assertions. As a general rule the chairman should be an intelligent layman rather than an expert. If he knows as much about

59

the subject as the other members of the group he'll find it hard not to join in.

Some producers invite the chairman to be a full participant. I've known this plan to work, but not very often. A participating chairman may abandon all pretence to impartiality; or in his anxiety not to take advantage of his position he may do less than justice to his own views. Occasionally producers try to do without a chairman altogether. The result is usually confusion, dominated by the speaker with the loudest voice and thickest hide.

In smaller stations producers frequently act as their own chairmen. I've had to do this myself. The system saves time, but that's about all you can say for it. Like any other form of programme, discussions benefit from the application of an outside ear, and brain.

THE REST

In choosing the rest of the group you'll naturally opt for people who understand the subject. But expertise isn't the only consideration. They must be able to express themselves effectively, and not averse to stating an opinion. As a producer I suffered dreadfully from thoughtful citizens who weighed their words and took an age to make up their minds. They may have been assets to society but they were death to a radio discussion.

As in a play, *vocal contrast* between the participants makes listening easier.

HOW TO DO IT

I recently took part in a broadcast discussion (on TV, as it happens). This was the sequence of events:
  (i) A producer telephones, states what is proposed and asks if I'd like to take part. I say I would.
  (ii) Two days later I get a friendly note recapitulating the main points of our conversation, telling me clearly when and where to turn up, and inviting me to let him know of any points I particularly want to raise in the discussion.
  (iii) I arrive at the studio centre at the appointed time (11.30

a.m.). Am taken to the hospitality room and introduced to my fellow discussers and the chairman. A drink is put into my hand. General conversation, punctuated by nervous laughter. Tension relaxes and we get on easy terms with one another.

(iv) At twelve o'clock we move down to the studio. There, sitting in a pool of light, with cameramen and others weaving dimly around us, we start talking about the subject of the programme. Everyone at cross-purposes. The chairman lets us ramble on for some time, then gradually begins to seize on points which strike him as of greater consequence than the others. The producer, booming away invisibly on talk-back, occasionally comes to his support. Order begins to emerge out of chaos. The chairman arranges the opening, in which he briefly introduces the subject and the discussers, giving each of us in turn a chance to say a sentence or two. We all rehearse this opening, then break off (just before one o'clock).

(v) Another drink. Producer says, to our surprise, everything going splendidly. Lunch (strictly teetotal).

(vi) After lunch, producer asks if we accept his list of important points to be covered. We do, apart from one or two extra ones, which he in turn gratefully accepts. He then suggests the *order* in which we should deal with them. This is logically faultless anyhow, so we again agree and file away to the studio area.

(vii) While we are being made up, the producer and chairman go into a huddle, to confirm, as far as I can make out, how much time each stage of the discussion should get. In the studio we again rehearse the opening, then go quickly through the main points of the discussion in the order agreed. Various new thoughts still come up, most of which the chairman welcomes.

(viii) At three o'clock the chairman tells us to relax, enjoy ourselves, and leave the rest to him. Shortly afterwards the recording begins. Opening goes swimmingly, and we all feel much more confident. Apart from noticing a floor manager occasionally tick-tacking to the chairman (no doubt indicating the time) I am conscious of very little but the flow of argument. New points crop up even at this

61

stage, but they are minor ones. All the major points are made as arranged, and suddenly it's all over.

(ix) Congratulations all round and an instant playback. We are given full details of the transmission date and told that cheques will be sent to us in due course. Tea, and away we go.

As an example of how to prepare for, and then run, a broadcast discussion this would be hard to beat. We were treated with the utmost consideration so that it was easy for us to give of our best. We were encouraged to speak our minds, yet dexterously and unobtrusively led. Radio producers, even of hurriedly-planned topical discussions, please note.

THE CHAIRMAN AND THE PRODUCER

The chairman of a radio discussion must keep on identifying the speakers by name until he judges that the listener knows which voice is which. But this he should do as unostentatiously as possible; otherwise everything gets slowed down. To have too many people talking at once is traditionally fatal: nevertheless you may take it that when the participants start cutting in on one another your discussion has really got off the ground.

I myself like to hear a chairman sum up at intervals and at the end. I don't mean that he should try to resolve all differences and supply compromise solutions, but that he should draw the threads together and indicate where there is agreement and where conflict. Not all producers feel as I do; many prefer to let discussions simply stop, or be faded out. But I suspect that listeners in general prefer to see things brought to some kind of conclusion.

Of all the duties of a chairman, one of the most important (and the most frequently neglected) is to see that *every* participant gets his fair share of air-time.

Live discussions are always a chancy form of broadcasting. So it's usually best to have one's discussion pre-recorded, to let it over-run slightly, and then to tidy it up for transmission. This procedure has the added advantage that at a pinch one can stop the proceedings and ask the chairman to start a particular section again – a desperate remedy but one it's sometimes convenient to use.

How does a producer keep in touch with his chairman during the discussion? By signs (but these can distract the participants), by notes taken into the studio (ditto), or by instructions given through the chairman's headphones (assuming he agrees to wear them). The last is probably the most efficient method provided you don't whisper in his ear too often. A fussy, interfering producer is a great nuisance to chairman and participants alike. On the other hand, the kind of producer who gets a discussion started and then goes into a coma is not, to my way of thinking, earning his money.

Most talks and interviews that get on the air now do so as items in magazines. All magazines, general or specialised, light or serious, long or short, are made up of separate bits and pieces, each complete in itself. (The self-sufficiency of the items is what distinguishes a magazine from the collage type of feature programme, with which it is sometimes confused.) Obviously, then, what matters is (a) the quality of the bits and pieces and (b) the skill with which they are put together. The splendid term 'editorial flair' means little more than the ability to choose good material and assemble it in the right order.

The assumption used to be that a magazine, like any other programme, would be listened to, if at all, from start to finish. In some cases the assumption is still valid; in most cases not. The newest kind of magazine, sometimes called a sequence, is very long, and deliberately aimed at the casual listener, the listener who leaves the radio on while she – or he – does the washing-up. (And why not? As Dr Johnson said of smoking, it preserves the mind from total vacuity.)

If you're put in charge of a magazine you should work teleologically, in other words start by defining your objective, and defining it as precisely as possible. If you do this you can hope to turn out a magazine with an attitude and idiom of its own. If you don't, the chances are you'll produce just another rag-bag. (Rag-bags have their merits of course, but I assume that what we have in mind is something more stylish.)

Your first decisions should dictate all the rest. They should dictate your choice of items and, most importantly, the particular angles you want your reporters and interviewers to concentrate on. They should dictate your choice of *compere*. They should dictate the amount of *music* you use. To ask, as some people do, how much music there should be in a radio magazine is like asking 'how long is a piece of string?' Your end should determine your means.

THE INGREDIENTS

Do remember how many radio resources are at your disposal. So often one hears magazines which are just one pre-recorded interview after another. This sort of monotony is unnecessary. Given up-to-date equipment, your compere should be able via the radio-car or the telephone system to do a *live* interview with anyone anywhere. You can use every kind of report, eye-witness account and short talk. Your magazine can include featurettes, discussions, news summaries, weather forecasts, programme trails, letters from listeners, items in verse, very short stories, even drama – I know of one magazine that carries a daily episode of a serial play lasting all of two minutes.

The featurette, or mini-documentary, is a particularly useful ingredient. It normally consists of first-person reportage plus actuality recordings. One of the best I ever heard was done for the British Forces Broadcasting Service, after the world-premiere in Berlin of Rolf Hochhuth's play *Soldiers*, which savaged Winston Churchill's record in World War II and in particular his policy of devastating German cities from the air. The featurette gave us a few moments of the play itself (in German) recorded in the theatre – very atmospheric – with a translation, then a synopsis of the plot, the reactions of three or four English-speaking Germans who'd been interviewed in the foyer, and finally a summing-up which touched on possible repercussions. All this in under five minutes, with perfect clarity and no sign of undue compression.

Featurettes, like full-scale documentaries, demand an eye for a subject, the knack of dovetailing actuality into narration, a certain originality of approach, and above all that lucidity of mind which enables a compiler to seize on what is most significant in a welter of facts. But featurettes can be put together very quickly: the BBC's Saturday afternoon sequence 'Sport on Two' often includes a few assembled and recorded during the transmission.

POLITICS AND POP

The last ingredient I want to mention, music in all its forms, deserves a section to itself. In the old days it was used *con brio*.

Apart from the mandatory signature tune it would come up remorselessly after every item, in a way that was supposed to be 'appropriate', sustaining the mood of the preceding item and modulating to that of the next. A nice idea, but it seldom came off. All too often the appropriateness of the music was confined to the title. After a piece about birds we might as easily get 'A Nightingale Sang in Berkeley Square' as 'Lo Here the Gentle Lark' with flute obbligato. The whole convention became a terrible bore – and a time-wasting one for the producer. I only mention it because in a few broadcasting circles it unaccountably lingers on.

Short magazines these days commonly carry no music at all unless it forms an integral part of a particular item – e.g. a featurette about amateur brass bandsmen or the centenary of a composer. Very occasionally, too, music is used in an evocative manner, thus:

SNATCH OF UNACCOMPANIED GAELIC SONG
PRESENTER: The music of the Scottish Highlands. Sixty-five years ago, today's guest, Sir Hector McNab, was born in a tiny village in the Outer Hebrides. . . .

But the present fashion in short magazines is to keep music to a minimum.

Some extended magazines, or sequences (the BBC's 'Today', for example) are equally austere. But in others (e.g. the local magazines of the British Forces Broadcasting Service) talks, interviews and featurettes float on a flood of light and popular music which has no bearing on any item in particular. Then there are magazines in which the music *contrasts* with the items. 'The Morning Show' on the African Service of the BBC uses Pop discs to hold together (or hold apart) serious, slightly conventional, talks and interviews of a mainly political nature. The effect is pleasing in a rum sort of way, and certainly distinctive. Finally there's the fashion for live music links in the form of topical verses 'sung to a small guitar'. This is fine if you like doggerel and third-rate folk singers.

VARIETY AND STRUCTURE

A magazine, we said, is made up of bits and pieces. If these are too alike the listener rapidly loses interest: editors and producers have to provide constant variety. Let's take a simple case. Imagine you're planning the order of items around a news summary, which is a fixed point. Other things being equal, the arrangement talk/interview/news summary/interview/talk is clearly more attractive than talk/talk/news summary/interview/interview. And what applies to type of item applies also to length and subject-matter. You have been landed, let us suppose, with two rather similar items. Where do you place them? Together? Yes, *if* there is some inevitable connection between them or if you can make a selling-point of the similarity. Otherwise, as far apart as possible. Even the tempo and the vocal quality of items must be borne in mind. If you are unlucky enough to have two slow speakers or two speakers with dull voices, don't juxtapose them if you can help it.

And remember that the music, if any, needs to be as varied as the words. Not so long ago I was listening to a magazine for British troops. Considering the scarcity of local originations the spoken material was remarkably diverse; yet the over-all impression was one of monotony. Why? Because, as I realised when we started to analyse the transmission, the girl producer had provided an overdose of 'her' kind of music – disc after disc after disc in the same idiom. This sort of mistake is easy to make, and fortunately easy to put right.

If yours is a magazine meant simply for dipping into, all you need to worry about is variety within the average listening-span, whatever that may be. In other cases you must pay attention to the structure of the programme as a whole. As usual, you need to open strongly, in order to compel attention. And, since you want people to listen to the next edition as well, you must try to close strongly, leaving a pleasant taste in their mouths. Some magazines start with a menu or contents bill – a list of the main items. A useful device but a two-edged weapon – for every ten listeners who 'stay tuned' because they like the sound of an item half-way down I dare say ten others immediately switch off because there's nothing on the list that attracts them, although menu-less they might have hung on in hope.

COMPERING

Magazines have always needed someone to introduce the items.

He used to be called the compere. Now he's the presenter, link-man, or anchor-man, each name indicating a different conception of his role and each role making a different set of demands on the producer.

The *presenter's* job is comparatively simple. All he has to do (apart from doubling as interviewer) is turn up on time and speak the lines which the producer puts in his mouth. When a presenter walks into the studio he can expect to be handed a complete script containing every word he has to say. The text may be altered during rehearsal, but seldom very extensively. This kind of compering is still much used in the BBC.

It's not easy for a producer to write continuity material that will sound natural when spoken by someone else. If you're forced to adopt this technique I suggest you remember these rules of thumb:

Write straightforwardly and concisely, while making sure that your words will arouse expectancy in the listener.

Keep things clear. If a reporter in a pre-recorded item says 'today' when the event occurred yesterday, account for the discrepancy beforehand. But don't make a meal of it.

Never try to *manufacture* a connection between successive items. If a real relationship exists, mention it by all means, but avoid old-fashioned cues of the 'From cricket we move to corned beef' variety.

Don't be heavy. Serious items can gain from flip introductions: the presentation doesn't have to be in unison with the items. But it ought to be in *harmony*, which means, for example, no hard-sells for off-beat material.

When the presentation is fully scripted it is sometimes possible to have a complete run-through, which is the best guarantee of a smooth show. If this can't be managed, do at least 'top and tail' the magazine: in other words, get the presenter to read out his introductions, and the studio managers to play in the beginning and end of the various tapes and discs on cue. You'll soon discover if your script makes demands on the SMs which are physically impossible – by no means a rare occurrence. The alternative is to make the discovery when you're doing the final recording or actually on the air – not recommended for producers with high blood pressure. In any case, it's hard on your presenter if he has to introduce material of which he's heard nothing.

Even if every item is pre-recorded, even if you know all the durations, even if you're convinced that there are no physical problems for the SMs, a full rehearsal can still pay off. It will give everyone that extra touch of confidence. And it will enable the presenter to do an even better job on transmission.

Some producers try to get the introductions, the running-order and the scripts of any live talks on to one master-document, the theory being that the fewer separate pieces of paper the panel operator has to deal with the better. The document includes every detail available 'at the time of going to press', with gaps in the typescript (to be filled in during rehearsal) for the rest. It might start off like this:

ANNOUNCEMENT (*from Continuity*): Station ident. MID-DAY MAGAZINE. Your Presenter is Bill Smith.
1. BILL: (Greetings) The Chelsea Flower Show is with us once again. What's the secret of its perennial success? Henry Jones has been talking to some of the exhibitors.
FLOWER SHOW TAPE (AX 7321) DURATION 1 min. 45 secs.
STARTS 'This year's show is bigger and better than ever.'
ENDS 'My advice to you is, don't miss it.'
1a. BILL: That was our reporter Henry Jones.

2. FANFARE (TAPE BX 8865, *from first leader*).
BILL: The trumpeters of the Plymouth Marines Band. This afternoon the President of Ruritania will receive the freedom of the City of Plymouth. James Robinson has recently returned from Ruritania, and he's with us in the studio to talk about some recent developments in that country, and about the importance of the President's visit to Britain. . . . James Robinson.
ROBINSON: (*live in studio*) Until recently, very few of us knew much about Ruritania. But during the last two years . . .

and so on. Given this sort of layout, changes in the running-order are child's play. All the director has to say is 'Switch items three and five', or whatever the case may be.

The charge most often levelled against fully-scripted presentation is that it makes a magazine stiff and muscle-bound. So it's worth noting that the BBC's 'Sport on Two', fast-moving and intensely topical, which draws on a vast variety of OBs and studio items, invariably uses this technique (with such ingenious features, mostly invented by the first editor, Angus Mackay, as three versions of each lead-in, to fit a win for team A, a win for team B, or a draw). What the sequence producers find is that the quicker the pace and the wider the selection of items the more essential it becomes to have a definite (but instantly adjustable) framework into which new developments can be fitted.

The *link-man*, at the other end of the compering spectrum, is expected to show a great deal of initiative. All he gets on arrival is a list of items, some sheets of background information, a list of possible questions for the interviewees and a pile of tapes and discs. How he sorts out these elements into a show is up to him (so long as his producer is happy). Any script he hands over for typing will be little more than a running-order, and his cues to the studio managers will mainly consist of hand-signs and talk-back. The result can be broadcasting of remarkable freshness and spontaneity. Equally, when the ad-libbing link-man has an off-day one gets waffle, vain repetition, and a series of introductions and back announcements that are slightly but wincingly off-key.

In local radio the link-man often 'drives the studio' as well. Sitting at something like a Continuity desk, he can not only put himself on the air and play discs, cassettes and tapes but mix-in telephone conversations and the output of the radio-car. And all this more or less at will, provided his technical assistants in the cubicle are on the ball.

Somewhere between the presenter and the link-man comes the virtuoso *anchor-man*, of the William Hardcastle breed, which was imported into Britain from the U.S.A. Unlike the link-man, who isn't hired as an expert on anything in particular – except, perhaps, broadcasting – and who often adopts a refreshingly semi-detached attitude to his material, an anchor-man is chosen as much for knowledge of the subject as for microphone personality. Flourishing for the most part in the Current Affairs area, he plays a creative role, actively influencing the choice and order of items and claiming a wide freedom to express his own opinions, both directly and by implication. Needless to say, he prepares his own links, but these tend to be fully scripted. (The custom is to have each lead-in typed on a separate sheet of paper – another method of ensuring that changes in the running-order shall be painless. Back-announcements tend to be improvised.)

70

This character has added a new dimension to British radio. He has also added to the risks of Current Affairs broadcasting, blurring the distinction between news and comment, and often appearing to commit a whole Public Service organisation to the views of one idiosyncratic journalist.

Some magazines use *two* comperes. If the transmission is extremely long, like the BBC's morning sequence 'Today', and if like 'Today' it exploits a lot of last-minute live material from different parts of the country, the arrangement has practical advantages. But they can be dearly bought. From the listener's point of view one compere is a necessity but two are a muddle – particularly when they are ad-libbers who oscillate unpredictably between straightforward introductions and cheery cross-talk. After all, when a compere says 'you' I take it he means me, the listener. It's confusing, and irritating, to find he's talking to someone else.

THE TOTAL IMPRESSION

Most items reach the studio in packaged form, i.e. on tape and fully edited. And such live items as occur tend to be put over with the minimum of rehearsal. So you might think that studio direction calls for little more than organising ability and technical expertise. But it isn't so. Ad-libbing link-men need to be listened to very carefully. But so, for different reasons, do presenters and anchor-men. And the *total impression* made by any magazine as it unfolds moment by moment needs attention: a collection of bits and pieces it may be, but it's also an entity. There should always be someone on duty concerned with more than timing and mechanics. The trouble is that so many topical magazines are not, in my sense of the word, produced at all. In some cubicles there's so much rowdy and unnecessary chit-chat, and there are so many assistants and hangers-on shuffling around and standing in front of the loud-speaker, that it's physically impossible for the man in charge to hear the transmission as a whole (assuming he wants to: the kind of unconverted newspaper man who now dominates BBC Current Affairs broadcasting usually doesn't). Hence those uncorrected mistakes in presentation and those ghastly lapses of taste (as when, after a mid-magazine news summary which is a catalogue of death and disaster, the anchor-man says brightly to the news-reader, '*Thank* you, Brian Martin'). And hence the boring shape – interview after interview after interview. My guess is that in these cases the producer or editor, concentrating as far as he can on the *content* of individual items, and half-deafened by the noise all round him, hasn't even noticed. If quiet in the cubicle cannot be attained I only wish that more magazine producers would follow the lead of Harry Walters, who runs that excellent mixture of music and topicality, the 'Jimmy Young Show'. Walters listens to each transmission alone in his office, from which there is a direct line to the studio.

Another gross fault is the abuse of the talk-back facility. Half the frenzied instructions given to comperes during trans-mission could be avoided if producers had done their preliminary planning more thoroughly. The presenter of 'Sport on Two' actually has to wear 'split cans', one headphone carrying the programme and the other a stream of talk-back from the cubicle; and the instructions don't stop even when he himself is speaking on the air. We are told that comperes don't mind – but I should like to hear their private opinions on the subject. After all, they are giving a performance on which their liveli-hoods depend. 'Sport on Two' may be a special case. In general I would affirm that it's quite unfair to worry your front-man with more chat during transmission than is strictly necessary.

Two further points. If you want to change the order of the items, or give any other general direction, speak concisely and make sure that *everyone* hears you. Secondly, if you find when you're on the air that the magazine is running short, don't lengthen the music (if you're using music) and don't tell any live contributors you may have in the studio to talk more slowly. I've heard both instructions being given, and quite idiotic they are. There is an optimum rate of talking in every case, and there is an optimum length for each particular sliver of music. A lively magazine that runs short is infinitely preferable to a dull one that doesn't.

EDITORIAL RESPONSIBILITY

There should never be any doubt as to who takes final decisions. In the BBC, at any rate, it's the producer or editor, not the compere, however expert or expensive, who is responsible for preserving a proper balance in matters of controversy. Ideas as to what constitutes proper balance vary with changing social attitudes (and changing Directors-General) but the *locus* of responsibility remains the same. It's worth remembering that the concept of balance applies not only to the material but to the way the anchor-man handles it. A couple of simple instances will illustrate the point. When, as often happens, coverage is limited to an interview with a supporter of one side, it's right that the anchor-man, if he does the questioning, should be tough, since in that situation he effectively represents not only the public but the opposition. On the other hand, when one side is interviewed and the other gets the greater freedom of a talk, fair play demands that the interviewee be fed the kind of questions that will enable him to say what he wants to.

KEEPING IT UP, OR THE ADMINISTRATIVE INFRASTRUCTURE

It's not difficult to produce one or two lively editions of a magazine. What counts is keeping it up, and making sure you don't run out of steam. You must have the machinery to provide a continuing influx of new voices, new ideas, and new pegs on which to hang old ideas. A popular magazine in a big station will have its own research assistants and a posse of freelances forever phoning in programme suggestions. In small stations the machinery may be no more than the producer's personal contacts plus a diary and a collection of running files and card indexes. If those are your circumstances you must make the best of them. You must keep your files and indexes up-to-date. You must *write down* every bright idea as it occurs to you and put the note where you can get at it when you are stumped for material. You must keep a list of likely speakers and the subjects they know about. Above all, you must keep a diary of coming events and relevant birthdays and anniversaries. This is all very tedious, but it's the only way to avoid the panic that assails a producer when he needs seven attractive items for the next day and can think of only three. Finally, do keep a full *ice-box* – a collection of items on tape which are not tied to any specific date.

LOCAL RADIO AGAIN

So far we have taken it for granted that the producer of a magazine or sequence does not himself take part. But in local radio he is often his own link-man, who also drives the studio. This arrangement may be necessary but heaven knows it's undesirable. No human being can simultaneously produce, speak the introductions, and (in drama terms) do the work of three studio managers. Or rather, no one can do all this adequately. Something has to give, and usually it's the element of production. Under such conditions standards are bound to slip. Those who deny it do a disservice to radio. Those who say its doesn't matter do a worse one. Or so I believe.

9: FEATURES AND DOCUMENTARIES

I

A certain frayed glamour still attaches to the term feature programme. In many parts of the world planners still believe that to dress up a subject, any subject, in the trappings of a feature – to use four voices where one would do and add actuality and music to taste – is somehow to guarantee not merely popular success but a bit of class. The planners are mistaken: they are confusing accidents with essence. But it shows how strong the old magic is.

The earliest features people were professional innovators. What they meant by a feature was any programme designed to exploit the potentialities and extend the range of the radio medium. For a long time, however, the emphasis on features has been on the *imaginative presentation of fact* – a sufficiently wide brief. A possible definition of the feature programme would be Fact as viewed by a creative radio man (or woman). The object is always to get beyond mere reporting and to involve the listener emotionally and intellectually. So a routine feature programme is almost a contradiction in terms. It's certainly a waste of money and resources.

II

DOCUMENTARIES

When features began in the BBC they were mostly studio-based, for the good reason that outside recording was so difficult. There wasn't much scope for spontaneity when your equipment had to be trundled about in a vast pantechnicon that got stuck under low bridges, and when you had to carve your material into four and a half minute chunks because four and a half minutes was all you could get on to a disc. The invention of electro-magnetic tape and the midget recorder transformed the radio scene, making possible a whole new category of programmes: the actuality feature, or documentary.

The terminology of the subject, I should add at this point, is confused. Until fairly recently *all* imaginative radio programmes based on fact were called features, a term which subdivided into dramatised features, actuality features and so on. But the radio feature was by origin a branch of the multi-art documentary movement. So dramatised features in television were called documentaries. These days BBC radio also uses the term 'documentary', but means by it 'actuality feature'. 'Feature' unqualified equals studio feature. The change is unhelpful. If one accepted it no word would be left to describe the *whole class* of imaginative radio programmes based on fact, or the various hybrid forms which *combine* actuality with studio material. So I can't accept it.

How is an actuality feature made? The first step, I suggest, is to make sure that actuality is the right technique to use. There's no point in recording simply because speakers are available. The question is, how much will their personalities and voices add to your programme? If the honest answer is 'very little', do consider if some other feature technique wouldn't do just as well.

It's worth remembering at this stage that actuality features don't *have* to be about social problems, or indeed crises or difficulties of any kind. It's quite enough if they increase our understanding of the world we live in or the people we live among, or if they merely show us that life is more interesting, entertaining and strange than we had previously suspected. A number of actuality features, for instance, compiled by George Ewart Evans and produced by David Thomson, were devoted to showing that in East Anglia and North-East Scotland old countrymen still held to secret beliefs about horses and horse-magic that connected them with Britain before the Romans. Valid feature material, surely. One of the most arresting actuality features ever broadcast was not about people at all but whales, and their 'song'.

Once you've fixed on your subject you rough out a plan of campaign – topics to be covered, areas to be visited, persons to be interviewed (I assume, for the sake of simplicity, that you are your own producer – not an ideal arrangement but one I've had to adopt many times). Don't get too set in your ideas at this stage. People who arrive on location with a ready-made scheme, stick to it and get away as quickly as possible, turn out programmes that are dead from the start. It's much better, time permitting, to let one line of enquiry suggest another, to let your plan ramify. On the other hand, you mustn't be carried away by the interest of the subject. My mainstay on these occasions is a paper-covered pocket-book in which I've listed the points on which I want information and the actualities I simply must get before I go home.

One uses the pocket-book for another purpose as well: to jot down further ideas, happy phrases, and brief descriptions of places and people – invaluable when it comes to composing narration. Years ago I went to Cyprus to compile a documentary on the latest crop of 'troubles', and the British people, civilian and military, who'd been caught up in them. A page in my notebook, which I recently came across, reads thus:

Pale stubble.
Eucalyptus trees – only green things.
Long sticky fingers.
Plain – colour of underdone toast.
Nicosia – orange stone. Pink roofs.
Lines of washing!!
NB. Ring Army PR chap before 6.15.
Walled city. Ataturk memorial? – staring white marble.
Old English queer in café.
Byzantine churches mouldering away (in countryside).

Many of these points went into the narration:

Cyprus, for an independent country, is very small – smaller than Yorkshire. It has two towering mountain ranges north and south, and a plain in the middle. At this time of year it's all barren rock and pale stubble – from the mountains it looks like a slice of underdone toast. Nicosia swelters down in the plain: it's about the size of Norwich, all orange stone and pink roofs, with lines of washing everywhere. The heart of Nicosia is the walled city, with one or two broad streets and crowds of narrow, rather smelly, medieval alleyways – a sniper's paradise. . . . When there's no shooting, and outside of high summer, Cyprus is a real picture-postcard zone, ideal for the retired Briton. One recalls its features: white sails gliding over the blue Mediterranean; vines, figs and olives; the long sticky fingers of the eucalyptus trees; the happy peasantry, still knowing their place and willing to work for next to nothing; dark Byzantine churches gently mouldering away; endless supplies of local wine; artistic English bachelors pursuing in exile their various foibles; garden parties for U.K. citizens. It was a kind of dream world that burst open last Christmas.

INTERVIEWING

The chief difference between interviewing for magazines and for actuality features is that in the case of features you ride your victims on a much looser rein. Most of these programmes are explorations, investigations in depth, and you can't dig deep in two and a half minutes. Nor will people say anything of real significance on cue. You need plenty of tape and a fine willingness to chuck it away.

Not long ago I heard an engrossing example of the actuality feature. It was about a man of 50, Joey Deacon, barely able to speak, living in hospital, who had somehow managed over the years to tell his life story to a fellow-patient, his great friend Ernie. Ernie, himself illiterate, had communicated the story to two other patients, who wrote it down. Now it's been published as a book. In 'Tongue-tied: the Silent Life of Joey Deacon', produced by Thena Heshel, the extracts from the book made wonderful listening, but what stood out equally strongly was the power of the interviewing by Nancy Wise. Her conversations with the hospital staff, with Joey's friends and with Joey himself, were unsentimental, understanding, infinitely unhurried. One shudders to think what the results would have been if the methods of magazine or news interviewing had been applied.

Again and again one feels on location that if one merely goes on recording, asking questions at intervals and not worrying about repetitions, false starts and long silences, something of real value may emerge. Occasionally nothing does. At other times this psychiatrist's couch technique yields rich dividends. I remember sitting in a Cyprus garden, looking at the dark smudge of the Kyrenia range as it melted into the evening sky, and just listening (with an occasional muttered comment) as an English housewife described the agony of having to decide every morning whether it was safe to send her children to school through the road-blocks. Came an exceptionally long pause, and then this story seemed to well up out of her:

The children had been asking for weeks, 'Please may we go to the mountains for a picnic?' They love it up there. We knew St Hilarion was occupied, so we went off towards the east. And we found a little track which seemed to lead towards the mountains. We went up this track and we went on and on for, I should say, three miles. Finally the road petered out at a village, and we got out of the car.

When I got out I saw that it was a deserted village. There was a contrast there. There were these very, very poor mud-brick houses – no plaster – and at one side, overlooking them proudly, on an eminence, was what remained of the village school. It was written – I could make it out: it was in Turkish – that it had been put up exactly a year before (I was there in March). It was a two-roomed place, and it had got plaster on the outside walls, and it had got windows, glazed windows – quite different from the ordinary, very poor, houses. And every building had had everything movable torn down or broken up, and fire-blackened holes where the windows had been.

The school had had special attention. It looked as if the pupils had been there in the middle of writing, and gone. There were – there were overturned chalks by the teacher's desk, all broken and crushed underfoot. The whole of the inside was littered with glass. There were exercise books with words half-finished, and desks and things as if they'd had hatchets at them. And it was *completely* silent. There wasn't a sound of a *thing*. . . . And outside, all round on this little hillock that the village was placed on, there was a *sea* of yellow daisies.

I haven't altered a word. Mrs der Partog really did say 'overlooking them proudly, on an eminence'. On the air that story, with its exquisite closing irrelevance about the sea of yellow daisies, was almost unbearably moving. And the point is, it would never have been told at all but for the atmosphere of sympathy which had somehow been established, an atmosphere in which she had almost forgotten the recording machine churning away in the darkness at our feet.

An ethical point strikes me here. One must never trade on that forgetfulness. If you suspect that in the cold light of dawn an interviewee might regret having said what he did, then the least you can do is play the tape back to him and let him know what it sounds like. You can forfeit good material by such fastidiousness, but less than you might suppose: the cost of honesty is surprisingly low. Besides, you might want to visit that part of the world again.

Don't be content to confine your recording to pre-arranged interviews and a few obvious sound-effects. I remember compiling a documentary about a cultural festival on the west coast of Ireland. The high-spot was not anything I'd set up with the organisers but a tape which it had suddenly occurred to me to make by the simple process of walking, recorder in hand, through the entrance-hall and the public rooms of my hotel. The voices speaking Irish and English, the country laughter in one bar and the Saturday evening fiddling and singing in another, together with the constantly shifting acoustic perspective, added up to something rather memorable. I admit that when you record off the cuff and on the sly you must be prepared for awkward moments. You also need to be aware of copyright implications and other legal niceties. But I'm certain that a touch of radio-vérité can enrich a feature immeasurably. People going about their business or pleasure, talking to one another rather than to you, often make the most evocative sound-effect of all.

A BBC producer, Peter Armstrong, has added a new dimension to actuality features. Going in his turn to Ireland, he induced a country priest to carry on his person a radio-microphone (i.e. a microphone plus mini-transmitter) during the whole of one day – an important day in the Church's calendar. Everything the old man said, to casual callers, to his young acolytes, to his congregation, was broadcast on low power, picked up by a receiver-van parked round the corner, and recorded. Unworried by the sight of an interviewer, the priest soon forgot to be self-conscious. The result was actuality of amazing impact and naturalness, full of rustic humour, unconscious pathos and spiritual beauty. 'Father Greene's Ash Wednesday Mass' was a clear step forward in feature technique – radio-vérité indeed.

Equally important, as far as I can judge, is the imaginative use which a West German documentary producer, Peter Leonhard Braun, is now making of natural sounds. Recorded by the most sophisticated means, they are often no mere additions to a narration: they can occupy the aural foreground for minutes at a time.

PROCESSING THE RAW MATERIAL

It's when you get home, tired, dirty and overspent, clutching reels of tape and a bulging notebook, that the most intellectually taxing part of the business begins. This is the stage that shows the difference between a features man and a hack. The hack is content to set out his recordings in 'large lumps', in roughly the same order as they happened to be made. He then supplies a few stereotyped links of the kind one can write in one's sleep: 'That was the manager, Mr Jones. I next had a word with a foreman, Tom Smith, who had this to say. . . . That was a fore-man, Tom Smith. When I asked one of the workmen, Bill Robinson, how long he'd been employed at the factory, he said . . . ' – and so on and so forth.

The real features man works very differently. He begins by *listening* to the actuality, and judging it with *rigour*. It's likely to fall into three categories: good stuff – to be used; moderate stuff – usable when edited; and poor stuff – not to be used at any price. It's important to classify entirely on merit: the trouble it may have cost you to obtain a particular recording is beside the point. Incidentally, if a piece of tape is unintelligible at first hearing, do remember that the listener will hear it only once. Playing it back in order to improve it by editing is fair enough. Playing it back until you've understood it, and *then leaving it in*, is idiotic. But I've seen it done. Having made a preliminary choice of actuality and gone over your notes, you next have the job of fusing a great many separate elements into a whole. When you went on location you knew which aspects of the subject interested you most. By now you'll have abandoned some and discovered others. Consider in detail the various points made – consciously or unconsciously – by your interviewees, and the new points that occurred to you as you went along. You'll become aware of *connections* between them – corrobora-tions, contradictions, contrasts. Bit by bit a better ground-plan can't fail to emerge.

At this stage it's useful to have a *transcript* of all the recordings you haven't rejected out of hand. BBC producers in London have transcripts made for them by a team of audio-typists. Working in a Region, I used to make my own. I would then take a pair of scissors, cut up the transcript and my notes, and pin together everything relating to point A, then everything relating to point B, and so on. This was a wearisome job, for my notes had been written down higgledy-piggledy as they came to mind, and the order of the recordings (any one of which might cover several points) had been dictated not by logic but by the availability of speakers. However, by the end of the cutting and pinning process I would have my raw material, their speech and my observations, arranged not according to the rough-and-ready headings I started with but on a new and (I hoped) artistically satisfying, developing, often dramatic, pattern. As Laurence Gilliam once said, 'an actuality feature is built up, formed, *composed* like a piece of music. It's the artistry with which the composition is achieved that distinguishes the feature from the mere compilation'.

THE IMPORTANCE OF NARRATION

Usually it's the narration, flowing around and between the bits of actuality, that holds a feature together and gives it unity. Well-written narration, interesting in itself, can take the place of the actuality one failed to get, lend an additional sparkle to good actuality, and make the rest sound better than it is.

I once did a 'Return Journey' feature on Israel, a country which I hadn't seen since it was Palestine. The programme started in this way.

ANNOUNCEMENT:	Elwyn Evans recently went to Israel, and in the programme that follows he gives you some of his impressions. . . . A LOOK AT ISRAEL.
EE (*Studio*):	It was a quick look at an unbelievable country. Everyone I spoke to seemed to have a life story that deserved a programme to itself.
EE (*Tape* 5):	How long have *you* been in Israel?
Girl:	Just one and a half year.
EE:	And where were you from originally?
GIRL:	I am coming from Rumania.
EE: (*Tape* 6):	And where did *you* come from?
MAN:	From Baghdad.
EE (*Tape* 20):	Where are *you* from?
MAN:	Montreal, Canada.
MAN:	South Africa.
GIRL (*Tape* 17):	I'm from Holland.

GIRL:	I'm from the United States – Chicago.
EE (*Tape* 8):	Where are *you* from?
MAN:	I'm from Cardiff, from Leckwith Road, near Ninian Park.
EE (*Studio*):	Yes, and he's by no means the only Israeli from Wales. I found half a dozen in Northern Galilee in a communal settlement, a *kibbutz*, just half a mile from the Syrian border. As we drove between the hills, with the Sea of Galilee trembling and glinting in the distance, past fields white with stubble in the strong sun, I didn't quite know what to expect. Barbed wire, no doubt, and sentries. What I did find, as we entered the totally unguarded gates, was a young father pushing a pram. Then lawns, flower beds and a very crowded swimming pool. . . .

Technically this extract is worth attention for several reasons. In the first place the montage ('From Rumania, From Baghdad . . .') shows how recorded actuality can be arranged. Slices from five different reels of tape have been juxtaposed. Secondly, you will have noticed that I went straight from studio narration to recorded question ('How long have *you* been in Israel?'). There's a school of thought that deprecates this practice, preferring the 'I asked' technique – thus:

EE (*Studio*):	Everyone I spoke to seemed to have a life story that deserved a programme to itself. I asked one girl how long *she'd* been in Israel.
GIRL (*Tape* 5):	Just one and a half year.

Personally I see no harm in cutting from Narrator in studio to Narrator on tape. To subject his voice to a sudden change of acoustic is refreshing to the ear. It also signals perfectly adequately that the programme is moving into 'an actuality situation'. One doesn't always have to spell things out. Thirdly, the extract shows, I hope, how narration can be used to give a programme colour, and thus heighten the effect of the actualities.

The finest compiler-narrator ever is René Cutforth. His pungent paragraphs, delivered in his peculiar grating voice, are models of their kind. Here's the beginning of his feature (produced by Francis Dillon) on one of the early Aldermaston Marches:

83

ACTUALITY OF SINGING AND SLOGANS

MAN (*Tape*):   This march is just a lot of bloody psychotics trying their best to divert attention from their own psychic difficulties, you know. But in the end to no purpose. It's like a bunch of tiny dogs yapping at the back door of a big house. It will accomplish sweet nothing.

WOMAN (*Tape*):   But I think if they've got the guts to show what they can do, good luck to them.

ACTUALITY: MARCHING FEET

CUTFORTH (*Studio*):   Between these extremes of bystanders' comments, the Aldermaston marchers . . . have been traipsing the trail to London for the fourth time. As it passed me into Maidenhead the column was about a mile and a half long, eight thousand strong, and it was a moving sight – not in the way of military marches but for precisely the opposite reasons. Though it kept perfect road discipline it was less of a march than a shuffle. The impact of its costume, in spite of the beatniks and other weirdies crowded up in front, was nil, except for the cameramen told off to make mock of the movement. The moving thing was the desperately sincere incompetence of these civilians as marchers; the good, sensitive faces, the presence of so many ordinary, worried people – most of them very young; more than half of them young women – shuffling along because, as one of them said, they have no platform but the public road, to call the British people's attention to an idea they have for avoiding the universal death promised by the H-bomb.

That's what an actuality feature should be like: a subject of human interest seen through one pair of eyes – intensely shrewd eyes. Cutforth's approach was strongly personal – rightly so: he was employed to give his own version of the facts, not the BBC's. But, though deeply involved in any subject he presents, he is, as he would put it, almost desperately objective and fair.

The best kind of actuality feature is a mosaic, with longer and shorter passages of narration and longer and shorter bits of actuality all so deftly joined together that they form a continuous whole. Make your joins as imperceptible as you can. Never hesitate to cut your own voice out of a recording. And never mechanically introduce every speaker by name: ask yourself, in the light of the extracts given above, if the name is essential (sometimes it is), whether a description of the speaker ('a student', 'a young soldier') wouldn't do instead, or whether even that is necessary. Take these words from my feature on the Troubles in Cyprus:

EE (*Studio*): In the meantime most English civilians press on, not exactly regardless but as far as possible un-involved, like well-bred guests at a party pretending not to notice.

MAN (*Tape*): So far as this present – er – situation is concerned I can only say that I've continued my ordinary life. I've been in and out of the Turkish Quarter and the Greek Quarter and I've never had the slightest trouble.

MAN 2 (*Tape*): There's one thing you can say about this country. It's the only country I've lived in where you can leave your car unlocked in the street, where you leave your house permanently unlocked. They're probably the most honest people I've ever known. Another thing is that the troubles here have had very little effect on your relations with your real friends.

It was clear from the context that these two were British civilians, and from their voices that they were middle-aged professional or business men. So why slow down the programme by introducing them? Why *not* let their voices float in completely unannounced?

A final remark about narration. It should never be a series of flat statements. It exists, among other reasons, to raise questions in the mind.

DETAILED TAPE-EDITING

When it comes to fine editing you can, if you have perseverance and a good ear, sometimes work miracles. Here's an example. An ebullient lady I came across in Israel who had English as a third, or possibly fourth, language, gave me this account of how Oriental immigrants were received on arrival:

The second problem is, people when they arrive here they . . . sometimes don't know the food they should know how to – er – they – the food suitable for them to our – er – climate and so on. So this people are also again, through voluntary groups and through social workers in the country, we give them to explain, to understand, the food which is – and that also the nurses help a lot and used to help before more; now it's less – er – because people are more understandable already – teach them the food they should know to eat here and which is suitable for the children. You know that many of the people here are right from Ira – from the Orient or say from Iraq or wherever it is, they . . . even – er – they have a lot of children. In Iraq one – er – so to say – one of the women once said to me, 'Oh, there th– a child died, all right, died, d'you see? Because we have the – God gave us other children.' Here they understand that we have to *save* the children; and – er – this women – we have to give them to unders- to know that the child has to be cleansed, washed, arranged, put to bed – and that's all work and problems which you have to deal with.

I badly needed Miss A's testimony. But to have broadcast her words as they stood would have exposed her, me and the programme to what Sir Thomas Browne called opprobrious scoffs. After listening to the tape many, many times I settled on what seemed feasible cuts and transpositions. These I marked on the typescript. Hours later, my tape editor and I surfaced with this revised version:

The second problem is, people when they arrive here they . . . sometimes don't know the food suitable to our – er – climate and so on. So through voluntary groups and through the social workers and also the nurses, we give them to understand the food they should eat here and which is suitable for the children. You know that many of the people from the Orient or say from Iraq or wherever it is, they . . . even – er – they have a lot of children. One of the women once said to me, 'Oh a child died, all right, died, d'you see? Because God gave us other children.' Here they understand that we have to *save* the children; and – er – we have to give them to know that the child has to be cleansed, washed, arranged, put to bed – and that's all problems which you have to deal with.

It made sense, and I guarantee that no one who heard the broadcast would have guessed that the tape had been doctored.

I could have smoothed out this lady's utterances a great deal more, but I chose not to do so. I left her sounding, as in real life, hesitant in her English, but no longer unbearably so. People have a right to their own personalities and their own tricks of speech, and cosmetic editing should never be carried so far as to turn them into different people altogether. Or to alter the *meaning* of what they say. A BBC example is apposite. Dr Ramsey, when Archbishop of Canterbury, was famous for his halting delivery. Asked to comment on the theological opinions of a fellow-cleric, he obliged, whereupon a well-meaning tape editor eliminated the hesitations and repetitions. The result was surprising. Dr Ramsey became much more easy on the ear: he also sounded about twenty times as definite. Cosmetic editing had had side-effects, converting diffident disagreement into outright condemnation, of a highly un-Anglican variety.

Respect for the evidence is crucial. When editing you abbreviate, you distil the essence of a point of view. To go further, to sharpen up attitudes in the interest of dramatic conflict – usually by omitting parentheses and qualifications – is unforgivable.

Some of the practical considerations to be borne in mind when you edit tape are these. First, you need to follow your *ear*. A transcript made by others is a useful aid, but even if it were totally reliable, which it seldom is, there's no sense in *starting*, as some do, by editing on the page. The *tune* of a sentence, the rhythm of a phrase, the precise location of a pause or a breath – all these matter a great deal. It should be apparent that they *define* the edits which are possible. It's equally true that they *suggest* edits which are often more effective than any you could devise by staring at a typescript.

It is, unfortunately, necessary to mention that you can fine-edit *only* if you physically cut the tape. Some broadcasting organisations, anxious to save money by using tapes over and over again, object to physical cutting and expect their producers to edit by dubbing. *It can't be done*, except in the crudest possible way.

When editing one should never trim a speech in such a way that identifiable sounds in the background which ought to fade away gradually (e.g. those of aeroplanes or cars) are abruptly cut off. If you do, you advertise the fact that editing has occurred, which is not desirable unless you are practising some new form of Brechtian alienation.

During the editing process don't throw away rejected pieces of tape in too carefree a manner. Stick the substantial ones on some convenient rail. If your programme starts to run short you'll then be spared the frustration of grubbing about on the floor in search of a particular edit which looks remarkably like fifty other edits, and discovering when you've played it through that it's not the one you wanted. A modicum of method is a considerable advantage on these occasions.

ASSEMBLING THE PROGRAMME

Before the final recording all the actualities should have been joined together, in their new order, to form a single reel, the start of each bit being indicated by leader tape.

It's even more necessary in the case of documentaries than in the case of magazines that the presenter/narrator should wear headphones and really *listen* to the recordings as they are played in. If he doesn't, the joins between actuality and studio speech will be anything but imperceptible and the programme will proceed in a series of learner-driver jerks.

Because the technical quality of studio speech is normally better than the quality of speech recorded on location, a narrator frequently *sounds* louder than the inserts even when the decibel counts are identical. So the panel operator should keep the level of the narration just a shade lower than that of the actuality. It all makes for smoothness.

JOINT-STOCK DOCUMENTARIES

Now and again one comes across a feature in which the duties of researching, reporting, interviewing and scripting have been divided between a number of people. This arrangement may be unavoidable when a producer is working against time on a big complex subject. Otherwise there's little to be said for it. Minor works of art, appealing strongly to the imagination, which is what features are supposed to be, are seldom engendered in committee. And I always feel that the hitherto uninvolved professional narrator who comes along with his voice beautiful or his ready-made reputation to bump up the audience figures adds the final touch of falsity to the whole proceedings.

NARRATIONLESS ACTUALITY

As soon as tape recording had got properly under way various BBC producers (notably W. R. Rodgers, Maurice Brown and

88

Denis Mitchell) began thinking in terms of 'pure actuality' –
programmes in which recorded voices, plus on occasion recorded sounds, should speak for themselves with the minimum
of narrative interpolation. Some of these 'tessellations of tapes'
put together in the late Forties and early Fifties were superb.
I still recall the intellectual excitement generated by the Irish
conversation-pieces of Bertie Rodgers, produced by Maurice
Brown, with their inspired cross-cutting, and the atmospheric
quality of Mitchell's 'The Talking Streets'. But the movement
petered out. Listeners obstinately 'wanted to know, you know'
– not all the time but certainly now and again – who was
speaking. On the whole they didn't care for voices in a
vacuum. Rodgers himself, whose 1949 feature on W. B. Yeats
had included next to no narration, began to use more and
more.

Producers were also becoming aware that in radio documentary as in radio drama the narrator didn't *have* to stand apart,
like an abortionist at a christening. In plays, narration and
dialogue could be 'deeply interfused'. In actuality features,
where the narrator had done the interviewing and the subject
was being looked at through his eyes, integration was easier
still. Here is a nice example from W. R. Rodgers' feature on
George Bernard Shaw (first broadcast in 1954):

NARRATOR: It was Mrs Shaw who, when Shaw was at
              a loss for a subject, suggested Joan of Arc.
              . . . Lord Glenavy recalls a certain night in
              Ireland when he and his wife and Shelah
              Richards were invited to the Shaws'
              sitting-room.
GLENAVY: When Mrs Shaw had seated us she said,
              'Mr Shaw thought he would like you to
              come tonight as he has just finished the
              last page of his new play'. . . . Then he
              gave us a long and fascinating talk, explain-
              ing how he had interpreted her acts and her
              life. . . . Then I heard my wife break out in
              a tone of bewilderment. 'But Mr Shaw,
              there's a lot that matters about Joan of Arc
              that you haven't put in!'. . . . He listened
              with twinkling eyes, as my wife mentioned
              instances of what she meant, until she
              suddenly gave up. 'You see,' said Mrs

	Shaw, summing up the discussion amiably, 'you have mainly felt about Joan. Mr Shaw has mainly thought about her.'
NARRATOR:	Shaw's great gift to the theatre was the play of ideas. In his plays, too, as in his life, there was that detachment from feeling, that figure-of-eight movement of the mind, which enabled him freely to take both sides of a question, to comprehend both saint and sinner. Perhaps it was his bi-partisan Irish background; but just as Joyce could sit on both sides of the sense, just as in Gaelic there are no words for 'yes' and 'no', so in Shaw's plays there is no plain white or black, no villains. All men are variations in search of a theme. And the music is in the counterpoint.
FATHER LEONARD:	One time he said to me that the first thing he thought of in casting players were the voices. He like to cast the play according to the type of voices, soprano, alto, tenor and bass.

(From *Irish Literary Portraits*, published by the BBC, 1972)

During recent years the concept of 'pure actuality' has been revived and extended. The leading revivalist – the name suits him – is Charles Parker, who preaches the gospel with vast energy and in a highly personalised style. Parker's great gift to radio is his pentecostal realisation that not merely picturesque peasants and wise old craftsmen but *everyone*, including urbanised conveyor-belters and the telly-debauched young, can speak, and that the distillation of what they have to say can be of vast importance, artistically and sociologically. Some of his narrationless documentaries have been stunning in their impact. It's the doctrinaire in him that every now and then spoils his work, making him in the interests of a theory seek to impose on his essentially impressionistic technique a weight of argumentation that it manifestly cannot bear. And forcing him into specious claims. He says he dispenses on principle with the narrator because narrators are 'the voice of authority', his recordings being presented 'in such a way that the listener [can] form his own conclusions'. But the kind of narrator Parker has in mind is a bogey invented by himself. And he forgets that by the very act of *selecting* voices, and subsequently patterning them so as to emphasise some and diminish others, he, Charles Parker, has predigested reality for us as effectively as any narrator, authoritarian or otherwise.

In Parker's documentaries narration was replaced by juxtaposition and reiteration. It is obviously much more difficult to convey any kind of paraphrasable message by such means; but a message, I am convinced, is what he often wanted to convey. Hence the very long time it took him to put his programmes together, and, by radio standards, their high cost.

All this is not to imply that there is no future for narrationless actuality, given a recognition of its inherent limitations. An experiment in this genre by Barry Bermange, 'The Dreams', in which accounts of dreams and nightmares were edited and rearranged into patterns of beauty and horror, has shown the continuing power of the technique and the possibilities of a new range of subject-matter.

It has become apparent that the main danger for the creator of narrationless actuality features is self-indulgence, usually manifested by undue length. This kind of programme is too often like certain kinds of late romantic music – all shifting surfaces and no construction, all tactics and no strategy. However deep the initial impression, however atmospheric and evocative the music or the feature may be, both become boring, cloying, if too prolonged. It's useful to remember that Samuel Beckett's latest plays, in which he dispenses with plot and even dialogue, are exceedingly brief.

COLLAGE COMPOSITIONS

This, I suppose, is the place to mention that in a series of twenty-six historical features called 'The Long March of Everyman' that clever producer Michael Mason applied some of Parker's methods to *written* sources. Documents of various kinds – letters, memoirs, poems, the records of legal proceedings – were minced up, the bits were scattered about, and key phrases were taken out of context and repeated over and over again. Narration was minimal. As an additional curiosity of broadcasting, the words were read out by an army of amateurs with local accents. This fake-actuality technique had some

91

stimulating results. It gave to statements made in court by persecuted nineteenth-century trade unionists, for instance, an extra dimension of pathos – though we were also landed with occasional absurdities, as when poetry written in mediaeval Welsh by ruling princes whose other languages were Norman-French and Latin was declaimed in the English of a Rhondda miner. Listeners who stayed the course must at least have gathered the impression that British history was crammed with human interest. Personally I rather wearied of being tossed about on a trackless ocean of sensibility without ever sighting an ordered argument.

In later compositions (his own term) Mason has gone a stage further, thickening ragouts of literary morsels with climactic chunks carved out of the standard musical repertoire. One's estimate of the value of such radio works must inevitably be related to the view one takes of the Collage movement in Art – and that view may well be related to one's age.

FOLK MUSIC LINKS

The use of specially-composed folk music to hold actuality together is a technique pioneered by Denis Mitchell and much developed by Charles Parker in his memorable series of 'Radio Ballads'. The device answered wonderfully in 'The Ballad of John Axon', the 'true life story' of a heroic train driver. I think this was partly because the musicians, led by Ewan MacColl, were building on a foundation of existing folk songs and ballads about railwaymen. The music was in key with the subject. It was equally apt to later documentaries on deep-sea fishermen and coal-miners. But its limitations were sharply exposed in a radio ballad devoted to sufferers whose lives are spent in iron lungs. To tragedy of such dimensions, plinky-plonk tunes and birthday-card poetry were a wholly inadequate response. However, this programme too was a technical triumph. One marvelled at the positively bird-like skill with which the briefest threads of music and speech were woven together. It's a tragedy that Charles Parker's restless brilliance is now lost to British radio. Some young producer really ought to take up where Parker and MacColl were forced to leave off.

92

III

Scripted features come in a variety of shapes and sizes. But (ignoring hybrid forms for the time being) there are two main categories: dramatised and non-dramatised. The second group we might call, more positively,

NARRATIVE FEATURES

In these a story is told or a subject analysed by a succession of studio voices without the aid of dialogue. Often a strong narrative thread is adorned with extracts from original sources. It's an austere form of radio, but one that can make compulsive listening.

A memorable example of the genre was a feature by Anthony Powell on the battle of Stalingrad. Using passages from despatches, war diaries and letters from the front, Powell movingly re-created the final anguish of Hitler's vast army, lost in the Russian snows, realising that they'd failed to take the city and that they themselves were now encircled. It was the extracts from letters – 'the last recorded thoughts of doomed men' – that provided the programme with its emotional peak. No invented dialogue, no music, no sound-effect was here: only the stark authenticity of the words. And the impression they left was overwhelming.

To success in compiling a feature like 'Stalingrad' there is a triple key. One must know how to select the passages that matter from out of the mass of original material, how to arrange them so that they make artistic and logical points and how to write narration that illuminates while it attracts. Production-wise, nothing is more important than matching voices to words. In 'Stalingrad' the various letters – from a regular officer, from a disillusioned ex-member of the Hitler Youth, from a military chaplain and so on – were each allotted to a different reader. The fact that every voice seemed exactly right added immeasurably to the poignancy of the programme. 'Characterised reading' has its victories no less renowned than acting.

Sometimes the interest in a narrative feature has been almost entirely intellectual, as in Isaac Deutscher's classic studies of 'The Great Purges' and other aspects of Communist rule in Russia, or, on a different level, the programme in which two doctors, Ida Macalpine and Richard Hunter, applied some recent discoveries of medical science to solve one of the puzzles of English history, the cause of the so-called insanity of King George III. At present, though, in the BBC at any rate, the narrative feature is chiefly used for scissors-and-sellotape radio-biographies of a fairly superficial kind. They make smooth, chocolate-coated broadcasting. I have produced them by the score, and written quite a few. Here is the opening of an un-usually good radio-biography, 'Eleanora Duse', by Thea Holme (from a series called 'Theatrical Portraits').

SHAW: A terrible thing happened to her. She began to blush; and in another moment was conscious of it, and the blush was slowly spreading and deepening until, after a few vain efforts to avert her face . . . without seeming to do so, she gave up and hid the blush in her hands.

NARRATOR: Bernard Shaw, describing a performance by Eleonora Duse. It's one of many attempts by many people to convey the astonishing simplicity and truth of her acting. Using little or no make-up, and speaking always in Italian, she didn't seem to be acting at all. She *was* the character. During this same London season in 1895, Gordon Craig, a young actor in Irving's company at the Lyceum, slipped across the road to Drury Lane in time for the last act of *La Dame aux Camélias*. He wrote to his mother, Ellen Terry:

CRAIG: I saw Miss Duse last night. Her wonderful and *god-*like death. I could not applaud – I could not think of Duse. I could only shake with tears and keep as quiet as I could in my seat.

NARRATOR: Some accounts, particularly of her later perform-ances, even suggest that at times she was possessed by a spiritual force from outside her. She herself wrote after her triumphant return to the stage in 1921:

DUSE: This success belonged to something far greater than I; far above my head; it was directed by a force which was not myself – I was merely its representa-tive.

94

Easy listening, without a doubt.

The narrative feature technique at its simplest demands little more than an eye for a subject (which every features man must have), the ability to write for the ear, a clear mind, and access to printed material. As such it's worth the attention of anyone operating on limited resources. But it *can* be used at a high level of imaginative intensity. Take Nesta Pain's celebrated series on bees, spiders, 'and such small deer'. Knowing no more of entomology than the next person, she read authors like Fabre, thought over their observations, and finally came up with material which could hardly have been fresher if she'd studied every activity through her personal microscope:

NARRATOR: The spider is an efficient huntress, but her efficiency is hardly likely to be a cause of satisfaction to her prey. They are obliged to evolve some means of escaping their destroyer.

NATURALIST: Yes – they must. For the spider population is so enormous that they are never far from danger – there is something like one spider to every three square inches of grassland. Some insects have found safety by growing a sort of armour-plating – the woodlouse or the beetle, for instance, draws in its legs and waits placidly for the attacking spider to realise that attempts to make a meal of it will only jar its fangs. Some insects have gone in for camouflage, but this is more use in protecting them from birds than spiders, as most spiders are too short-sighted to notice subtleties of colouring. They don't use their eyes much in hunting except to see movement. And there are certain kinds of flies and moths which have developed a flavour that is highly unpleasant to spiders. These resort to a very desperate device to elude death.

MUSIC BEGINS.

VOICE: Flutter through the sunshine and enjoy yourself – if you can. This field, this little plot, is the limit of your world – and it is a world which holds two million spiders in a single acre, two million jaws waiting to seize you, if you stray their way. It's better to fly than to crawl. But death hides its snares even in the air. Take care, as you fly, of the almost invisible web, the faint gleam in the sunshine. . . . There! You've blundered into it! Your

95

wings are sticky, your feet held fast; the more you struggle the more helpless you become. It's everybody's nightmare come true – the pursuer close upon you and your feet sinking into the morass. The pursuer at your back and you can't run away. *You can't!* It's like trying to escape in soft snow from a pursuer on skis. . . . You don't see your enemy? No – not yet. But her hand was on the skein that leads to her web and now the message has come that you're there – each plunge, each despairing, frantic struggle for freedom sends the message travelling along the line, the message for which she was waiting. . . . And here she comes! Gliding over the web that holds you helpless – a flowing grace of movement, accomplished almost before it is seen. You may well struggle in your panic and despair as the monster comes near, each leg longer than your whole body, her swollen belly bigger than your whole bulk. . . . *But don't!* Don't struggle! Lie still – it's your only chance. Let her bite you once! You may survive that savage, poisoned bite. But if you struggle, she'll bite you till you're dead. She's coming close now. She's watching you with those eight eyes of hers that shine like jewels, though they see little enough. Lie still . . . quiet . . . still . . . the moment's coming now – the bite's coming. Bear it – bear this one bite. It's your only chance. . . . Now!

CHORD. MUSIC ENDS.

NARRATOR: The spider bites and the insect lies still. There is no struggle, no resistance. The spider, revolted by the taste, throws the insect to the ground, and then staggers clumsily to the side of the web and is sick.

The real stroke of genius here is the sudden switch from third person to second – the address to the struggling fly. Anthropomorphic, yes: but how effective!

CHANGES OF VOICE

Note, too, the neatness of the handovers. So often in narrative features the text seems to be split up arbitrarily, as though the author had thrown his paragraphs to the readers like sweets at a children's party:

With one for him and one for he,
And one for you and one for ye,
And one for thou and one for thee.

Here every change of voice makes sense because it coincides with a change of intellectual focus. It's dreadful to hear the producer of a narrative feature, faced with a long speech, announce without a moment's hesitation, or thought, that it must be 'broken up'. A good passage of narration, complete in itself and all of a piece, is better left unfractured. And if the passage is *not* good, giving a few lines to one narrator and a few to another doesn't help. Vocal variety won't disguise rotten writing. Far more useful for the prevention of monotony than change of voice is *change of tempo*. Experienced writers of narrative features know how to build a script out of separate sequences so that a change of tempo between one sequence and the next becomes almost inevitable.

IV

DRAMATISED FEATURES

Many dramatised scripts of the past – 'Under Milk Wood', 'Alfie Elkins', 'Hilda Tablet' – went on the air as features mainly because the BBC's drama experts were then uninterested in work that was considered to be 'loosely constructed', 'formless', or 'lacking in dramatic tension'. Features producers on the other hand, briefed to promote creative writing and innocent of formal dramatic training, knew no better than to encourage such authors as Dylan Thomas, Bill Naughton and Henry Reed. Ideas as to what constitutes a play have undergone a revolution, and all these productions would now be unhesitatingly classified as drama. This is a retrospective triumph for the old-time features brigade but it also means that the scope of dramatised features is much reduced. By and large they are now confined, like narrative and actuality features, to the imaginative presentation of fact. (And even when so confined they are now sometimes called plays. A recent case in point was David Rudkin's 'Cries from Casement as his Bones are Brought to Dublin'. Billed as a commissioned play, it was an old-fashioned studio feature if ever I heard one – and, of course, brilliant.)

How effective are dramatised features for such a purpose? It's evident that actuality is often unobtainable – for instance, when you're tackling a historical subject or a recent episode behind the Iron Curtain. It's also evident that a story told in dialogue will usually be livelier than an account by one voice or relays of readers. So there will always be a market for dramatised features on such themes. *Radio Times* is full of them. But is this the only use for dramatisation – to project a subject when you can't get actuality? I don't think so. I once produced a dramatised feature by Alan White on the impact of short time and redundancy on industrial South Wales. The next day the BBC broadcast an actuality documentary by Brian Blake on short time and redundancy in the North of England. This is what Ian Rodger wrote in *The Guardian* about the two programmes. Alan White's feature

moved without connecting narrative from one short realistic scene to another. It listened mostly to the frightened talk of two steelworkers faced with redundancy, but though their characters were sufficiently established to have made them fit subjects for a play, Mr White had a larger canvas in mind. He introduced Lloyd Hopkin and Joe Davies, but he also brought in a meeting of the town council, discussions in the boardroom of the steelworks, and even eavesdropped on chemists arguing about the use of automation in the making of steel. He thus achieved a total picture of the effect of short time and redundancy on his community. With a tape recorder he could have interviewed similar people, but though he might have gained in authenticity of detail he would not have conveyed the larger truth.

To achieve point [the feature] has to aim at archetypes and it is very difficult to find archetypes in the street. Mr Blake's subject was the same as Mr White's but he was using the method of the tape recorder enquiry. He interviewed men on the dole, union officials and employers, but though their stories were tragic they remained personal experiences rather than archetypal ones. The perspective obtained by Mr White's 'Redundant' eluded him. His human material could not condense experience with the sharp economy of the voices from Mr White's informed imagination.

Brian Blake was at least as good a producer as I, and if my programme made a deeper impression it simply proves that there are times when the dramatised feature *form* is the best one to use, whether actuality is available or not.

As striking a dramatised feature as I can remember was a

study of life in a women's prison, 'Twelve Months, Mrs Brown', written by Kathleen Smith and produced by Dorothy Baker. The opening was constructed thus:

Judge pronounces sentence. Cross-fade to interior monologue ('Twelve *months*? In *here*?') counterpointed by sound of footsteps along stone corridor. Footsteps stop; monologue breaks off. Sound of cell door unlocked: opened. Dialogue, Prison Officer/Mrs Brown. Door locked.
Short scene: evening meal.
(Mrs Brown alone in cell.) Interior monologue: Mrs Brown breaks down: cries herself to sleep.
Bang on door: morning sounds.

A feature consisting of the recorded testimony (if obtainable) of prisoners, warders, the Governor and so forth might have given us some discomforting insights into prison conditions. What it couldn't do is enable us, in our imaginations, to *live* the experience of one 'archetypal' prisoner. An actuality feature might have described (in the past tense) the miseries of a first night in gaol. That would have been disturbing, but far less so than what creative writing alone can supply: the sense of undergoing the experience with the prisoner *now*. The justification for dramatised features on contemporary themes is that invented dialogue in the mouths of invented characters sometimes expresses the essence of a human situation better than any words you can dig out of real people.

What goes for social studies goes for much else. Many of us have compiled 'Return Journey' programmes on documentary lines. But by no other means than dramatisation – narration plus dialogue – could Dylan Thomas, in his 'Return Journey to Swansea', have shown us the whole range of his feelings as he approached the blitzed town, and have united the voices of the present with those of the past. When one tries to evaluate the potential of the dramatised feature it becomes clear that autobiography is as important as biography – witness many fine programmes, for instance 'The Streets of Pompeii', Dorothy Baker's 'Return to the Black Country', and that marvellous series 'A Year to Remember'. Then there were all those dramatised character studies, 'District Nurse', 'The Rugby Club Secretary', and the tragi-comic pictures of community life supplied by Bill Naughton and Gwyn Thomas (who that heard them can forget their contrasting accounts of small-town trips to the seaside?). And what about those Imaginary Conversations? – the remarks Bonhoeffer might have made to Niemoeller about theology; what Robespierre might have said to Wordsworth, or Pasternak to an English reporter, about the theory of Revolution. The scope of the form is so much wider than most contemporary producers (and Service Controllers) realise.

FEATURES IN AFRICA

We have been talking in terms of a sophisticated audience. It's worth noting how strong the impact of dramatisation can be on the young and the unsophisticated. You can hang anything on a story-line. When I was concerned with broadcasting to the illiterate people of rural Nigeria I found that incomparably the best way of explaining the alleged advantages of modern farming methods was by using 'coarse works of art' – *dramatising* the ill-success of conservative peasants and the prosperity of their neighbours who had followed the advice of the Agricultural Officer. It was a hundred times more effective than putting the Agricultural Officer on the air.

THE RESPONSIBILITIES OF THE WRITER

For 'a true and fair view' of a contemporary situation listeners to a dramatised feature depend entirely on one person, the writer. Critics tend to dwell on this difficulty as though it had no parallel in other forms of the feature. But it isn't so. The writer's responsibilities are no heavier than those of the compiler of a documentary, who also has to select from among a jumble of facts and opinions those which are the most salient. Another accusation is that sloppy writers can dream the whole thing up at home, dodging the all-important leg-work. This is true. But it's equally true that sloppy compilers can cook up some sort of documentary on site, dodging the all-important brain-work. The moral is not 'Away with the dramatised feature on contemporary themes' but 'Away with sloppy craftsmen'.

It might be helpful if I listed the three most familiar weaknesses in dramatised feature scripts. They are, I think, as follows:

100

(i) *Cardboard characterisation.* Ian Rodger wrote of archetypes. Yes, but it's fatal if the characters are *merely* typical. They must be drawn, or at least sketched, in the round.

(ii) *Bogus dialogue.* Most dramatised features are naturalistic, or meant to be so. Writers whose main interest lies in a social *situation* often don't worry enough about truth to nature in their characters' language. Or they may not even *know* how some of their fellow humans talk. It's a chastening experience to re-read old dramatised feature scripts and notice what passed for naturalistic dialogue in the days before portable tape-recording. Even now there's room for improvement.

(iii) *Dramatising the undramatic.* It's when a writer tries to force a set of reluctant facts into dialogue that the dramatised feature really breaks down. The radio playwright (*vide seq.*) faces a similar problem. In either case, why disdain narration? All one has to make certain of is that listeners sense an organic connection between the narrative passages and the rest. No one can enjoy the sort of feature in which all the intellectual interest, all the development of the theme, occurs in the narration, on to which little scenes in forgetable dialogue are hung like presents on a Christmas tree.

V

DIRECTING STUDIO FEATURES

The direction of actors in a feature poses certain problems which don't apply to plays.

In the case of literary or narrative features the *degree* of characterisation appropriate to 'characterised reading' matters. It varies from programme to programme; but in general one may say that too little characterisation leads to dullness while too much (as when an actor goes beyond *suggesting* a personality and reads a document as though he were delivering a speech) turns the programme into a farce. It is for the producer, not the cast, to decide what is required.

In dramatised features the scenes tend to be short, and also

101

independent of one another, being connected only by narration, so that actors often have to make an impression in a very few lines, starting from cold. The inevitable result is that they press hard on the accelerator. Furthermore, actors can be amazingly ignorant of life as it is lived away from theatres. Having little or no conception, for instance, of what judges and barristers sound like, they are content to reproduce in court-room scene after court-room scene the same well-loved caricatures of the legal voice. Most dramatised features, as I've pointed out, being naturalistically written, the first requirement is that one's cast should sound like 'real people'. A producer usually has to spend a great deal of time and energy persuading his actors not to act – in the pejorative sense.

Since there is little development of character in the average feature, since actors have to represent *given* characters in mini-situations, the producer, unlike the producer of a play, can very seldom afford to sit back and let his cast work things out for themselves. During rehearsals he has to supply much more of the dynamic than his colleague, and go in for much more detailed direction. This applies particularly to the argument, or logical thread, of a feature, which not all accomplished actors are good at grasping. There's a special kind of delivery – rather loud, rather fast, rather 'thrown away' – accompanied by a set of all-purpose inflections, which actors always employ when they don't quite understand the significance of a speech. Be not deceived. Take the speech – not the actor – apart, and work at it with him until both you and he are happy.

Casting, it should be remembered, is relevant not only to the effectiveness of one's production but to the intentions of the writer. In 'Twelve Months, Mrs Brown' the part of the first Prison Officer was played by an actress with liquid tones and a culti-vated accent. The use of a different voice, older, harsher, would without the change of a single word have altered the whole picture presented to us – rendering it, as it happens, a good deal more credible.

Finally, it's worth noting that precisely because dramatised features tend to be written naturalistically and to include short scenes they offer scope to non-professional actors who haven't much technique, or much ability to sustain a part, but who do have natural talent. As a former Regional producer I know well what splendid earthy qualities such actors can bring to a programme. This is an important point for producers in developing countries.

MIXED MODES

Our division of features into three main categories was only rough. There's plenty of overlapping, and some thoroughly mongrel features have proved far livelier than the pure-bred variety.

Narrative features quite often include nuggets of dialogue. A half-hearted use of dramatisation, perhaps, but it can be attractive. After all, the real words of real people have already been uttered, and uttered in character: it's only a short step to making the various personages address *one another*. When the original documents contain plenty of direct speech the dialogue almost writes itself. All you have to do is cross out 'he said', 'she rejoined', and so forth: what you are left with is dialogue. (This isn't a new technique. See Boswell's *Life of Johnson*.)

Another hybrid is the *narrative feature which includes actuality*, an interesting form, hideously over-used by the World Service of the BBC. The trouble is that its very able staff writers are expected, for purely administrative and financial reasons, to churn out features at a rate more appropriate to the composition of leading articles, and that the quickest way to give a subject some degree of synthetic featurisation is to write an essay, distribute it between our old friends Narrator I, Narrator II, Reader and Expert; then trick it out with a bit of music (on gramophone records) and actuality (provided by a freelance). Good features can't be written to a formula, and they are essentially personal – the subject, as we've said, must be viewed through one pair of eyes, not through the Cerberus-gaze of an institution. Nevertheless, the fact that this particular form is 'soiled by all ignoble use' should not blind us to its potentialities.

A feature I've already mentioned, 'Tongue-Tied: the Silent Life of Joey Deacon', consisted mainly of actuality but also contained narration, readings (from Joey's autobiography) and a discussion. Because Nancy Wise was interviewer, narrator,

chairman of the discussion and compiler of the whole pro-
gramme, unity of tone was beautifully preserved. It was so with
David Rudkin's programme on Roger Casement. As Anthony
Thwaite wrote in *The Listener*,

Rudkin built his complex mosaic in an impressionistic way, making
rapid transitions in time, place, space and manner, but he never lost
the essential narrative thrust forward, even when, towards the middle
of the programme, he stepped away from Casement and went into a
quick trot, almost in news-flash style, through Irish history. Dramatised episodes, passages from the diaries, bits of what purported to
be recordings of Rudkin himself addressing a Belfast literary society,
Joan Bakewell interviewing a miraculously revived Casement (now
an ancient peer), a pageant-play procession of ghost-voices from
Irish history, the final homiletic dialogue with an Ulsterman: all
these potentially disastrous ploys were managed with marvellous
skill.

Skill is certainly what you need if you intend to mix your
modes. Its absence ruined a recent feature on the origins of the
Second World War. This consisted mainly of 'characterised
reading'; without descending to mimicry the actors sounded
sufficiently like Churchill, Chamberlain and the rest to make us
feel that we were somehow in their presence: disbelief was
willingly suspended. Then suddenly, about a third of the way in,
we heard the *actual* voice of Neville Chamberlain, recorded at
the beginning of the War. The effect was catastrophic. When the
actors resumed they sounded like a bunch of phoneys; the
whole mood of the programme had evaporated. Yet I guarantee
that if the recording had been more adroitly placed, say towards
the end (or even at the beginning, separated from the readings
by a chunk of narration) it would have worked beautifully.

What of the corresponding difficulty, that of sliding a little
dramatisation into a documentary? Another feature on World
War II (they are coming thick and fast these days) introduced
us to various old men reminiscing about the Battle of Britain
and Air Chief Marshal Dowding. All quite riveting, until we
suddenly heard words by Winston Churchill which had been
uttered thirty years before – not recorded but imitated, and
imitated all too well, by someone on the Rep. Authenticity
disappeared, and with it all conviction.

A feature, like any other piece of radio, is sounds coming out of a box. It's the imagination of the listener that gives them life. As in the case of radio drama, that imagination is usually equal to the most rapid leaps in thought *within a given frame of reference*. But arbitrary alterations to the frame of reference itself cause trouble. A listener's approach to 'real people' describing what they've actually experienced *must* be different from his approach to actors speaking invented dialogue. Both modes of communication are valid, but they belong to separate *orders* of validity. Radio is an artificial medium and largely an affair of conventions. A convention established at the beginning of a programme ought not to be changed en route without the most anxious consideration. I labour this point precisely because within the basic artificiality of a single-sense medium the range of options open to a features man is so wide – so wide that utter shapelessness, justifying all the old sneers, becomes his constant temptation – hence the all-too-familiar kind of feature which has music in the first part and none in the second, unexpected spasms of pointless dialogue, random sound-effects, bits of actuality in otherwise subjective sequences, and personal and impersonal narration all mixed up together. Eventually the listener sighs for the simple joys of a straight talk.

And yet it's possible to maintain a programme's self-consistency, however various the ingredients. I revert to that early steelworks feature. The building of this vast structure made an extraordinary story – too extraordinary, I felt, to be told in the conventional fashion, with recordings of workmen, designers and directors. So I started the programme in this way:

MACHINERY SOUNDS UP LOUDLY AND SUDDENLY.
THEN DOWN TO BACKGROUND.

NARRATOR: This is a story of Power: the story of how an immense new Works arose by sweat and magic out of a wet expanse of sea-marsh and sand – a Works which will pour a torrent, a flood, of steel into a steel-hungry world – steel from Wales.

UP SUDDENLY A MALE CHOIR: 'HARLECH'.

NARRATOR: Those were steelworkers, steelworkers of Wales. I know them well. My name is Richard Burton, and I was born amongst them, in a crowded valley on the sea coast of Glamorgan that leans confidentially towards the West Country, across the Bristol Channel. Once, on the endless after-noons of my summer holidays, I used to leave the grey streets of Port Talbot that lie criss-cross along the narrow plain between the mountains and the sea; I would turn my back on the ice-cream cafés and the unemployed shop windows and the level crossing, and make for the sandhills and the pools and the spiked grass of Margam Moors, that roaring green and yellow playground of ours where echoless boys' voices reached no further than we could throw a pebble before the sea-wind whirled our words away.

SEA-WIND HAS BEEN CREEPING IN.

FIRST BOY: Dick! Dick – oh!
SECOND BOY: What d'you want?
FIRST BOY: Look!
SECOND BOY: Where?
FIRST BOY: By there! Over the sandhills!
SECOND BOY: (*indrawn breath*)
NARRATOR: The wild geese. Snatching up my heart as they flew arrow-like under the sun, further and further inland, over the sandhills, over the main road, over the ruins of the Cistercian abbey and across the sudden hills. Those were my yesterdays on Margam Moors. And today? Today I look at the Abbey Steelworks, like a huge stationary ship anchored forever at the edge of the sea.

Over-written, I dare say, by current standards. But I'm sure that the idea behind the opening was right. The subject had a streak of wild poetry in it which simply had to be brought out. And the male singers had struck the note of actuality as well as art in the first minute. So it became possible for further sequences of narration and dramatisation to co-exist with such down-to-earth actuality as this:

When I arrived here the site was a lot different to what it is today. Nothing but lakes and marshes. There was so much water that we couldn't use lorries, only using horses, and when they were going across the moors it was a sight to see 'em there, as if the horses didn't have no legs and the carts didn't have no wheels.

106

The entire feature moved on two different planes of art and actuality, of subjective and objective, right to the end.

VII

The feature form is inexhaustible. Its subject-matter is the whole of observed reality, and its technical possibilities are unlimited – or only as limited as the wit, imagination and literary power of its creators (plus, let me add, the facilities and finance available: you can't do outstanding features on the cheap). It's good to notice that after a period of decline the form is now showing signs of new life – and not only in Britain. To end this chapter, let me repeat the two pieces of advice which I so often heard Laurence Gilliam give when addressing new producers in BBC Staff Training sessions.

First, *always go to the sources*, or as near them as you can get. Let's assume that you are writing a radio-biography. Is it really necessary, you may ask, to wade through the Life and Letters of your subject, in three volumes, with cross-references to other 'damned thick books', when some boiled-down version for schools is conveniently to hand? Yes, it is. What the listener wants is your personal view of this individual. Unless you get to know him as intimately as possible, all you'll be able to offer is a set of variations on someone else's theme. On a more practical level, you will almost certainly find that the three-decker biography contains all kinds of entertaining anecdotes, perfect for radio, which are missing from the briefer studies. To distil what has been distilled already is all very well for whisky manufacturers: it doesn't work for feature writers. The same point applies to actuality. I suppose I could have put together an adequate steelworks programme by recording the Managing Director, the Chairman, the architect and a few tame nominated workmen, linking the interviews with a smooth rewrite of the publicity handouts which the Information Officer was only too ready to supply. In fact, I spent many hours splashing through Glamorgan mud to survey the work in progress, and, in the most literal sense, to get the feel of it. I also considerably extended the Company's short list of interviewees. Some of the more alienated types, who told me in lurid detail what it was like to work in cold weather amid the girders a hundred feet up, contributed a note of reality to the programme which the Management's favourite sons could never have supplied. The malcontents were living, breathing, original sources.

Secondly, *try to be honest*. It isn't always easy. Features are persuasive, and it's a historical fact that the BBC's Features Department reached its first peak when it was engaged in pouring out War propaganda – honest propaganda, so the producers hoped, but propaganda all the same. Today features men in most African states work under the surveillance if not the direction of Ministers of Information. In all countries freedom of enquiry and expression is, and always has been, restricted by the 'implicit major premise' that the existing form of society must be maintained. The BBC prides itself on its objectivity and its exemption from Government control, but when one surveys the long vista of its features and documentaries it's instructive to notice how often problems that urgently needed ventilation were ignored in a genteel conspiracy of silence, disguised as good manners and the avoidance of unnecessary controversy. Nor, it seems to me, has the situation fundamentally changed, in spite of appearances to the contrary. There's a case for holding that the vast extension of sexual reportage and fantasy on the air of Britain serves mainly to divert the people's attention from the double locks that still stand between them and some of the facts that matter most.

There are difficulties within as well as without. When a documentary maker is too much a child of his time – when, in the jargon of the sociologists, his unconscious inferential structure corresponds more or less completely with received opinion – he acts as his own censor and removes all dangerous thoughts from his programme without any real awareness of what he's doing. At the moment a major threat to intellectual honesty in radio comes from high-minded and well-educated young producers, many of whom, like the rest of their age-group in the universities, regard detachment and objectivity not as ideals to be aimed at but as so much outmoded bourgeois crap. The great features men held very different views. They believed in the existence of objective truth, and tried in their programmes to get as near it as possible, regardless of institutional convenience, fashionable assumptions and their own prejudices. Some of these people were poets, others prose-writers. Some were colourful conversationalists. Lots were drunks. But when it came to the assessing and reporting of fact they were all as austere as accountants.

10: PLAYS

More has been written about radio drama than about any other form of radio. In Britain, as in many Commonwealth countries, hundreds of original radio plays are transmitted every year. So their main characteristics ought to be pretty widely understood.

They all stem from the fact that the audience can't see the actors, or the setting. This peculiarity is at once the basis of the form's appeal and its toughest limitation. Radio plays are less intrinsically compulsive than stage plays, television plays and motion pictures. And invisibility presents the radio dramatist with severe practical difficulties – how to convey physical action, for instance. On the other hand there are characteristics of radio drama which work to his advantage, giving him opportunities that scarcely exist elsewhere.

'IT'S ALL IN THE MIND'

He can afford, for instance, an extreme fluidity of construction. The audience for radio drama, as for all forms of radio, is an audience of one (indefinitely repeated). As one person listens to disembodied voices and their accompanying sounds, his imagination inevitably begins to work on the material provided. Since there's nothing to look at, the action has to take place exclusively in the listener's mind. And the mind can cope with the most rapid changes from place to place, from period to period, and from internal monologue to realistic dialogue and back again. The flexibility of the radio play is fully equal to that of the novel.

Precisely because the listener can't see the characters he visualises them – and the physical setting as well. To no other kind of drama does the audience contribute so much: listening to a radio play is itself a creative act. As the BBC's Drama Script Editor says, the glory of the form is that it's incomplete. 'No beautiful woman cast as Helen of Troy in the theatre would

110

please everybody. But each radio listener can construct a dream girl according to his own specifications.' The main function of radio drama is to suggest: writers need never pile on more descriptive detail than is necessary to activate a listener's imagination.

The visualising process can be extended to create characters and settings that would defy actualisation on stage or screen. As W. H. Auden expresses it, 'the disembodied voices of radio . . . can present such things convincingly, for the imagination of the listener is not spoiled by any collision with visual reality'. Thanks to radio drama's fantasticating power, coupled with its widespread use of internal monologue, or soliloquy (the revival of which has been one of its liberating gifts to drama in general), subjectivism flourishes on the air. We've all heard those pieces in which someone articulates (usually in monologue) his fear, disorientation and guilt, while we wonder with increasing uncertainty which snatches of dialogue are 'real', which are the expression of his neuroses, and which are both. Such plays, as Irving Wardle points out, exemplify one of the main upheavals in modern European literature: the emergence of individual consciousness as the only certainty in a world of flux.

I'm referring, of course, to the sharp end of radio drama. Most radio plays are totally devoid of ambiguity and *Angst*. But even these, however conventional their characterisation, are able to move with equal nimbleness through space and time.

Flexibility of construction has its limits, of course. A listener's imagination gets to work more easily on presented order than presented chaos. A switchback technique, employed throughout a full-length play, leads to nothing but listener's vertigo: dramatists nearly always provide some sustained scenes. Many good radio plays (and some masterpieces, like Beckett's 'All that Fall') include nothing else.

A radio play is built up of scenes as a television play is of shots. It helps to hold the listener's attention if the scenes vary in length, in location (as I've mentioned, a change of acoustic background refreshes the ear), in the pace of the dialogue and in the number of characters included. The last point is worth stressing. Admittedly, for reasons we'll come to shortly, radio drama is not favourable to ensemble-playing; but there's no reason except dramatist's funk why radio plays should consist of an uninterrupted succession of two- and three-handed scenes. It pays to open things up once in a while with some aural equivalent of the wide-angled shot.

'ONLY CONNECT'

Although each scene is distinct from the one preceding and the one following it, the action shouldn't advance spasmodically, like an amateur movie. A great deal of skill can be expended on unobtrusively preparing the listener's mind in scene A for what he is to hear in scene B, or scene C. Middlebrow drama in particular demands suspense and a strong line of development: scenes need to be linked together, not merely butted together.

THE SNAGS

Now for some of the disadvantages of radio from the playwright's point of view. First comes the need to work on a reduced scale. For reasons that have to do with the very nature of the medium, a radio play has to be shorter than a stage play or feature film. Various celebrated radio-dramatic works, especially those of the post-war era – 'Christopher Columbus' or 'The Rescue', for example – may be cited as exceptions to the rule, but it looks as if they will remain exceptions. (Irving Wardle rather nastily described them as 'self-conscious cultural gestures: officially protected products of the poetic drama movement' which lingered on in radio 'long after the cult had run its brief course in the theatre'.)

Secondly, there's the need to maintain a continuous high level of interest. Most stage plays have their quiet passages, sometimes prolonged, whose object is to make the crises and climaxes when they arrive all the more telling. But a radio dramatist, knowing, like a radio speaker, how quickly listeners can be lost, dare not let the interest sag below a certain point. So dramatic contrast becomes harder to achieve.

Thirdly, there's the need to keep down the number of characters in a given scene. The ear of the average urbanised listener is simply not sharp enough to distinguish between more than a few voices at a time – and even these need to be differentiated whenever possible by age, sex, accent or individual tricks of speech.

112

(Admittedly there are some writers – Don Haworth, for instance, and his predecessor Gwyn Thomas – who care for none of these things. Every Haworth character talks like every other Haworth character, and every Thomas character like Gwyn Thomas. But marvellously funny though the results are, it must be admitted that the dialogue of both writers is hardly ever self-sufficient: each depends heavily on narration.)

An actor on the stage can maintain total silence for long periods and then join in the dialogue with no trouble at all. He's been *observable* all the while, and not infrequently most eloquent when he had the fewest lines. But a radio audience can only be aware of a character's presence *when he says something*, or has just said something, or is being directly addressed. In every other case he starts to drift out of the mind's eye. Even in two-handed scenes, if A's speech to B goes on too long we start to wonder if B is still there. (This is one reason why radio players resort to 'vocal reaction' – those wordless murmurs and broken interjections which come from a character addressed in a long speech. They are certainly a substitute for the play of facial expression, and for gesture, but quite often their main purpose is to remind the audience that the second character hasn't gone off yet.)

What it boils down to is that in radio ensemble-playing is next to impossible. Scenes have to be confined to a small number of characters (five is about the maximum), with occasional crowd scenes by way of relief. And once a character of any significance has been brought before us, he must either be disposed of by a definite exit or provided with lines right to the end of the scene – which is a nuisance when his remarks are superfluous. The alternative is worse – to let him gradually dematerialise, thus giving the proceedings an air of fuzziness and unreality.

INFORMATION THROUGH DIALOGUE

The listener's imagination must have *some* descriptive detail to build on. So one test of a radio dramatist's technical skill is whether he can give *in his dialogue* a whole range of information over and above what characterisation and plot require. The information relates to physical action, to the setting, to the appearance of the characters – in short to what audiences in the theatre or cinema can see for themselves. It has to be carried without apparent effort, or else the dialogue becomes ludicrous. The classic example of the all-too-explicit line is

I warn you! This gun that I'm carrying in my right hand is loaded.

But there are plenty of others that could be quoted, including

Gosh, Mary, that blue dress with white buttons you're wearing looks super with the dark red handbag!

Even such masters of the medium as Giles Cooper slip up now and again; witness a passage from 'Mathry Beacon':

RITA: Listen to that Jake Olim. Can't half play it.
BETSY: What d'you talk to him about when you're alone?
BLICK: This and that.
RITA: Always on about Jake, she is. Gone on him, I reckon.
BETSY: Ooh I'm not!
RITA: Yes you are. Shove over. I want to sit on the doorstep too.

No beginner could have written a more unnaturally explicit line.

Given patience and ingenuity it's usually possible to construct speeches which both sound appropriate in the mouth of character A and also carry information as to the appearance of character B.

Don't look so frightened, man!

is a perfect example.

One can often make dialogue more natural by paring down one's first draft. Here's an extract from an excellent radio play by Elizabeth Holford about a rich old lady, her paid companion, and a tough nurse:

NURSE: Lucky I came in. She was getting out of bed if you please. If I hadn't seen her she'd have fallen again. We shan't be able to leave her alone.
COMPANION: That's absurd. We can't be with her all the time. I'll tell you what to do. Move the chair round (*sound of her doing so*) so that it's against the bed. . . . Now she can't get out.
OLD LADY: I don't want the chair there.

Not bad; but the companion seems to be spelling things out for the audience's benefit rather than the nurse's. In rehearsal we made a cut so that the last two speeches ran:

114

COMPANION: That's absurd. We can't be with her all the time.
I'll tell you what to do . . . (*business: movement, creaking*) Now she can't get out.
OLD LADY: I don't want the chair there.

The sound-effect indicated that something was being moved. Two seconds later the old lady's speech told us what it was. Altogether neater, you'll agree.

This brings me to another point which seems to come as a painful surprise to each new generation of radio dramatists. It is that sound-effects on their own usually signify very little. We do indeed instantly recognise the sound of a telephone bell, of doors opening and shutting, of birds. Most other effects, however, we can't identify until we are *told* what they are. For this state of things there are many reasons. Not only have we lost the countryman's sharpness of hearing; even the finest electronic reproduction of sound is different, in quality and often in volume, from the real thing; and again, whereas in ordinary life our identification of sounds is assisted by a whole set of indicators ranging from smell to place and the time of day, these are not available to us when we listen to radio drama.

THOSE SEAGULLS

It was discovered years ago that even the sound of a calm sea, which one would have thought basic enough, wasn't recognised by listeners. Hence the addition of seagulls crying: sea and sea-gulls together do the trick. But very few sounds can thus be improved into recognisability. Unaided, most effects can't set a scene: still less can they convey physical action. But once associate them with dialogue and they start to mean something. In an interesting mutuality the lines explain the effect and the effect adds richness and impact to the lines.

I've often watched this phenomenon being demonstrated at BBC seminars for radio writers. They listen to the recording of a large, rather, quiet, crowd and are asked to guess what it represents. The variety of answers is always astonishing – 'a supermarket', a 'a railway terminal', 'a London club at tea-time'. Then the recording is played as it was used on the air, thus:

FADE IN CROWD NOISES AND HOLD FOR A FEW
    MOMENTS.

TELEPHONE RINGS: RECEIVER LIFTED.

GIRL: St Andrew's Hospital. Can I help you?

Instantly, but instantly, it becomes apparent that we are in the reception hall (a large one, as we can tell by the acoustic) of a presumably important hospital. And instantly the muffled voices and the occasional footsteps begin to form a contrasting background to the girl's impersonal tones, investing the scene with depth and a certain diffused suffering.

The sound-effect and the explanatory line of dialogue don't have to occur simultaneously: they needn't even be close to each other. A well-tried way of making a wordless but perfectly clear change of setting is to 'plant' the necessary information beforehand. Characters A and B agree to meet 'on Friday afternoon, at Waterloo'. The play moves on: some time later (after an intervening scene, perhaps) we get the sudden sounds of a shunting train, footsteps, and an amplified station announcement. And we react as sharply as Dr Pavlov's ill-used dogs: 'Of course. It's Waterloo Station, on Friday afternoon.'

Donald McWhinnie described one Giles Cooper play as 'radio at its virtuoso best: forty-five minutes of highly distilled experience crystallised into a sound-complex; words, rhythms, evocative noises, fused into a kind of music score which constantly stimulates the ear and the imagination' (Cooper: *Six Plays for Radio*, published by the BBC). It's the producer's business, not the author's, to make sure that sound-effects are used so as to give listeners the richest possible reward. But it's for the author to lay down guidelines. Sometimes – not nearly often enough – a sound-effect is central to the play. In 'Mathry Beacon', Cooper's comic, prophetic and upsetting fantasy about a detachment of soldiers, men and women, who, under pretence of obeying orders, quietly drift out the War and then out of civilisation as practised in Britain between 1945 and 1954 – in this play the weird sound of the pointless lump of machinery the detachment is supposed to be guarding, the Watling Deflector, dominates the action; and the use of the sound is carefully and exactly prescribed by the author. Similarly, in 'The Waste Disposal Unit', Brigid Brophy's hilarious black comedy, the second half revolves around the 'quiet but deliberate champ-champ' which rises to an immense climax of clunking and gurgling as

116

the appalling American matron who owns the unit falls in, to be utterly consumed. In the average radio play, however, effects are only intermittently important. When the sound of a telephone bell or a knock at the door, or whatever, really counts, the dramatist should write in precise directions. But there's no need to litter a script with such instructions as (*footsteps*). The purpose of effects is to make plays more convincing, or more atmospheric, or funnier, or just clearer. When they don't subserve some such artistic end they are superfluous, whether they correspond to reality or not.

ACOUSTICS AND PERSPECTIVES

Acoustics can be as evocative as sound-effects, if not more so. Why should such matters be so often left entirely to the producer? It would be refreshing if playwrights paid far more attention than they do to the acoustic background of particular scenes. To perspectives also. Intelligent script editors would welcome, as signs of an imaginative grasp of the medium, such directions as (*in a large, empty room*) or (*speaking from an immense distance*). It would make a nice change from (*footsteps*).

NARRATION

'The scene is a large garden. Arthur and Edwina are strolling along a winding path that leads to the shrubbery.' This is the sort of utterance that gives the narrator a bad name (in no way improved by an alias, such as Storyteller or Chronicler). The presence in a play of a third-person narrator standing well outside the action is an almost infallible symptom of failure to think in radio-dramatic terms, and in particular of inability to convey information through dialogue. In the early days of radio this kind of narrator was a hideously familiar figure, but he's now confined for the most part to that popular genre, the half-hearted or half-baked adaptation, either of stage plays, in which the original dialogue is left almost untouched, or of classic novels, as the BBC calls them, in which all the bits in direct speech are given to actors and all the rest to the narrator – dramatisation made easy.

The *creative* use of narration, and particularly of first-person

117

narration, thoroughly integrated with the action, is a very different matter. Admittedly, it enables the playwright to dodge various technical difficulties, but it can give rise to others: and in any case radio drama is not an obstacle race in which every writer has to go through the same hoops. Nor does the fact that narrators, whether integrated or otherwise, are little used in theatrical, film and television plays have much bearing on the very different, single-sense, medium of radio drama. To suppose, as some theatrically-minded radio producers still do, that narration ought *in general* to be avoided is an over-simplification which takes no account of the fact that the subjective approach of radio drama is one of its major strengths. W. H. Auden writes in his introduction to *Persons from Porlock and Other Plays for Radio* by Louis MacNeice that the most striking difference between radio drama and the ordinary stage play is that in the former *everything* the characters say is primarily a soliloquy. 'In a stage play the audience "overhear", so to speak, the remarks which the actors they see address to each other; in a radio play each remark is heard as addressed directly to the listener, and its effect upon the invisible characters in the play is secondary.' This is, I think, a considerable exaggeration, but the element of truth in it goes far to explain why the sort of soliloquy known as first-person narration is so suitable to the radio medium.

Sometimes the integrated narrator earns his corn by the sheer power and beauty of the words he has to speak. Take the opening of 'Gazooka', the Gwyn Thomas play that recalled the greatest coal strike in British history:

FAINTLY, DRUMS AND GAZOOKAS PLAYING 'SWANEE'.
THE SOUND APPROACHES GROWS LOUDER, THEN
FADES AWAY.

NARRATOR: And to my ears, whenever that tune is played, the brave ghosts march again and my eyes are full of the wonder they knew in the months of that long idle sunlit summer of 1926. By the beginning of June the hills were bulging with a clearer loveliness than they'd ever had before. No smoke rose from the great chimneys to write messages on the sky that saddened and puzzled the minds of the young. The endless journeys of coal-trams on the inclines, loaded on the upward run and empty on the down,

118

ceased to rattle through the night and mark our dreams. The parade of nailed boots on the pavements at dawn fell silent, and day after glorious day came up over hills that had been restored by a quirk of social conflict to the calm they had lost a hundred years before. When the school holidays came we took to the mountain tops, joining the liberated pit ponies among the ferns on the broad plateaus. That was the picture for us who were young. For our fathers and mothers there was the enclosing fence of hinted fears, fear of hunger, fear of defeat. And then, out of the quietness and the golden light, partly to ease their fret, a new excitement was born. The carnivals and the jazz bands.

Rapture can sprout in the oddest places and it certainly sprouted then and there. We formed bands by the dozen, great lumps of beauty and precision, a hundred men and more in each, blowing out their song as they marched up and down the valleys amazing and deafening us all. Their instruments were gazookas, with an occasional drum. Gazookas; small tin zeppelins through which you hummed the tune as loudly as possible. Each band was done up in the uniform of some remote character never before seen in Meadow Prospect; Foreign Legionnaires, Chinamen, Carabinieri, Grenadiers, Gauchos – or what we thought these performers looked like. There was even one group of lads living up on the cold slopes of Mynydd Coch who did themselves up as Eskimos, but they were liquidated because even Matthew Sewell the Sotto, our leading maestro and musical adviser, couldn't think up a suitable theme song for boys dressed as delegates from the Arctic.

And with the bands came the fierce disputes inseparable from any attempt to promote a little beauty on this planet, the too-hasty crowding of chilled men around its small precious flame. The thinkers of Meadow Prospect, a small and anxious fringe, gathered in the Discussion Group at the Library and Institute to consider this new marvel. I can see the room now, and hear their voices. Gomer Gough, the chairman, broad, wise, enduring and tolerant as our own scarred hillsides, sitting at his table beneath two pictures, a photograph of Keir Hardie and an impression, done in charcoal and a brooding spirit, of the betrayal and death of

119

Llewelyn the Last. Then there was my uncle, Edwin Pugh, called Pugh the Pang for his way of wincing at every mention of the bruises sustained by our species in the cause of being so special. Then there was Milton Nicholas. . . .

It was on a Tuesday evening that Milton took my uncle Edwin and me down to an emergency meeting of the Discussion Group.

The long passages of sustained narration that Gwyn Thomas composed for me when he began working for radio, glorious though they still sound, now have an outmoded air. For the time being fashion has moved towards a slimmed-down style, in which narration glides and flashes trout-like through the stream of dialogue. Of this style Don Haworth is a master. Here is a typical passage from his play 'We All Come to It in the End'. George, a young man, is working as a Father Christmas in the toy department of a vast store.

CREEP IN CAROLS AND TOY DEPARTMENT SOUNDS.

GEORGE: Now, son, how old are you?
BOY: Six.
GEORGE: Six, are you?
BOY: Seven.
GEORGE: Are you six or seven then?
BOY: Don't know.
GEORGE: (*narrating*) It sounds ridiculous but I enjoyed the Father Christmas part of the business while it lasted and there was only one incident that you might regard as untoward.
MILLIE: (*worried*) There's a young lady asking for you, George.
GEORGE: Who?
MILLIE: She said Miriam.
GEORGE: Can I leave the Grotto?
MILLIE: Just for a minute while we're slack. Father Christmas will be back in a minute, children.
GEORGE: (*narrating*) Miriam was waiting by the teddy bears. I was surprised by her asking for me because I'd knocked about with her a bit now and then but we'd both got bored and it fizzled out before the Christmas season started.
MIRIAM: It suits you that Father Christmas outfit, George.
GEORGE: What do you want to see me about, Miriam?
MIRIAM: Can't we go somewhere private?
GEORGE: No. They court-martial Father Christmases that desert their post. What is it?

120

MIRIAM:	I don't know how it happened.
GEORGE:	What happened?
MIRIAM:	I think I'm pregnant.
GEORGE:	(*narrating*) 'Course birds always say that if it's fizzled out and there's nobody to loaf around with, but she really was blubbering and it presented a ridiculous sight, her blubbering and me in my Father Christmas outfit.

(From *We All Come to It in the End and Other Plays for Radio*, published by the BBC.)

Some radio dramatists have shown considerable ingenuity in devising disguises for their narrators. In 'The Prisoner', Don Haworth's blackish comedy about Stanley Warburton, who confessed to trying to murder his mother, the first-person narration takes the form of a statement by the accused read out by a policeman in court, to which we return between flashbacks. As a rule Giles Cooper dispensed with the services of a narrator, but when he used one, as in 'The Disagreeable Oyster', he cut him in half – a typically brilliant stroke – and gave us, so to speak, Ego and Id:

BUNDY:	I'm sitting at my desk on a fine May morning, wondering whether it's worth starting anything else before the week-end begins.

DOOR OPENS NOISILY

GUNN:	Bundy! Good man, Bundy; glad you're still here.
BUNDY:	Yes, Mr Gunn?
GUNN:	Bundy, there's a crisis, pin your ears back and listen.
BUNDY MINOR:	Mr Gunn has ginger hair growing out of *his* ears.
GUNN:	E.C.W.'s Stoddeshunt Works have just rung through.

(From *Six Plays for Radio*.)

A WRITER'S MEDIUM

We have noticed that the dialogue of a radio play often has to carry a weight of information which in other dramatic media is conveyed visually. But radio drama is a writer's medium in a much wider sense than that. Elsewhere the old primacy of the text is under heavy attack. As John Barber put it in one of his *Daily Telegraph* articles:

There has been a tendency in recent years to reduce the importance of the spoken word in drama. Playwrights have discovered that once they get away from people just talking in a room, whole new worlds of expression open up.

The trend has been stimulated by the feeling that words in themselves are not what they were: they have become shabby, imprecise and mendacious. As tools of communication, they are blunted not only because we all misuse them every day, but because politicians and advertisers and professional persuaders regularly poison the wells of language.

Not so long ago dramatists relied on words and little else. The plays of Bernard Shaw, of Ibsen, of O'Neill, are endless gabfests. John Osborne is almost the only modern writer of distinction with the same trust in the dictionary. Even so, the long speeches in his plays, piling ever more outrageous similes on top of ever more excoriating metaphors, often sound like language at the end of its tether. Harold Pinter uses words sparingly. He even has a play called 'Silence'.

One respected theorist, Antonin Artaud, urges the modern producer not to consider any text as definitive or sacred. He asks dramatists to try to recover for the theatre the notion of a kind of unique language, half-way between gesture and thought. Hence much contemporary mangling of classical plays.

And John Russell Taylor, in *Plays and Players*, says:

'Experiment' in the theatre, deliberate innovation courting unpopularity and incomprehensibility, has for the most part moved to other areas than dramatic writing: to instant theatre, improvisation and the 'happening', to the work of self-consciously exploratory, innovating directors and designers. Though one or two of the very young, very new dramatists like Heathcote Williams and Howard Brenton seem out, on occasion, to prove exceptions to the rule, on the whole written drama, verbal drama, as opposed to improvised and non-verbal theatre, is almost by definition, however outlandish its style, taken to be Establishment, on the side of the squares.

But radio drama still stands or falls by the word, and the written word at that.

It's true that *improvised* radio plays are regularly broadcast outside Europe. I've often seen them being put together in Nigeria by illiterate but highly-skilled casts working in Ibo and Hausa. And improvised radio drama has commanded large audiences in India. It does seem extraordinary that in Britain, with its crowds of young actors and actresses all trained in the necessary techniques, and with every facility for pre-recording and tape-editing, the BBC should have initiated no experiments whatever, as far as I know, in the improvising of radio plays. Of course the experiments would be expensive to mount, and all improvised plays, if they were to be of an acceptable standard, would probably cost more than the conventional anecdote-in-dialogue that can be knocked off – read-through, rehearsal and recording – in half a day. So, since British radio is likely to be dominated for the foreseeable future by jackbooted cost accountants, its actors will in all probability continue to read their parts off pieces of paper.

In any event it seems certain that the text, whether scripted or unscripted, will always be paramount in radio drama. Other elements contribute to the 'sound-complex': 'effects', acoustic backgrounds, music in all its manifestations; but they have seldom been allowed more than strictly subordinate status: in this respect 'Mathry Beacon', 'The Dark Tower', in the closing passages of which Britten's music takes definite artistic precedence of MacNeice's words, and perhaps a very few other plays, remain exceptions to the general rule – triumphant experiments that were never properly followed up.

It may be, of course, that by making words so important radio drama is avant-gardeish as well as old-fashioned. Tastes in the theatre don't last for ever, and some critics are already expressing the view that there's a limit to the amount of strobe lighting, amateur juggling and full-frontal nudity that theatre audiences will accept as substitutes for a strong text. But that's by the way. The new developments in the theatre have taught us, or rather reminded us, that, to quote John Barber once more, 'our dark intestinal urges are just as valid and human as the civilised instincts which are satisfied by the cerebral drama of reason and ratiocination'. The question for radio dramatists is, can these urges be put into sounds, music and (chiefly) words? And if so, how? The fashion in radio writing, we know, has moved from the florid and overwrought to a sparer, frequently witty and allusive, style, the style of which Giles Cooper was, and Don Haworth is, such a master. Its great merit is that it plays even better than it reads. Essentially it is conversation, tightened, sharpened, made (often in the most literal sense)

123

killingly funny; the phrases, the interruptions, the pauses all chosen and subtly counterpointed to make contact with the listener on a disconcerting variety of levels. This kind of apparently naturalistic writing often achieves its impact by methods as artificial as Noël Coward's. It has had a marvellous run. For that very reason, radio dramatists ought now to be feeling around for new means of expression. Some dramatists are.

11: DRAMA PRODUCTION

When the time arrives for a playwright's ideas to be realised in transmission the producer becomes the key figure. It may be that in the theatre and the cinema the current cult of the director is overdone; but in radio drama the predominance of the producer is unavoidable. For this there are several reasons. A radio-drama producer tends to be his own impresario. As often as not it's he who's chosen the play, and it's always he who chooses the performers. Such publicity and material reward as radio can still command are his to bestow. More important, during the production process he becomes a Supremo willy-nilly. Actors need an audience, and he *is* the audience. Not the ultimate one, of course, but the only one with whom a cast can be in direct contact. Like Everest, he is *there*; more than either a film or a television director he is constantly available, a piece of human litmus paper responding to stimuli, able to indicate which performances are succeeding and which aren't. His every facial expression counts, as witness the glances the cast keep darting at him through the glass partition. Secondly, among those present only the producer is fully aware of all the elements in the 'sound-complex'. The actors can't be certain of hearing anything except what goes on in their own studio, and this may bear little resemblance to what, after electronic treatment, emerges from the loud-speaker. As for the technical crew, one of them sits with the cast and hears no more than they do; the second, whenever he proposes to play-in tape or disc, has to shut his ears to stretches of dialogue, while the third, at the mixing panel, is too occupied with the mechanics of building the aural structure to appreciate its developing proportions. So a great deal hinges on the taste and judgement of the producer. By his preliminary planning and his flow of off-the-cuff verdicts it's he who moulds and forms the whole transmission. No radio play can be more spirited or elegant or inventive than the producer allows it to be.

125

But, however much actors and crew depend on him he in turn depends on them – not merely to carry out his wishes but to provide their own ideas and their own dynamic. Production is A.C., not D.C.

THE FIRST STEP

What are the essentials of drama production (many of which have a bearing on the production of dramatised features, dramatised programmes for schools, dramatised light entertainment and so on)? The first step, obviously, is to acquire a script. BBC producers enjoy the help of a central Script Unit, which receives a torrent of unsolicited material and has plays, suitably edited, available on demand. Even so, most producers also rummage around for themselves – a point worth noticing by their counterparts in some developing countries, hampered by traditions of respect for authority and the still-powerful Colonial habit of running radio as part of the Civil Service. Effective radio can't exist without some exercise of choice at producer level.

Having acquired a decent script you have to *study* it, in an effort to discover the author's intentions – a more difficult task than it used to be, as Charles Lefeaux remarks. You must ask yourself what the characters are like, where the element of conflict lies (is it physical conflict, the conflict of ideas, or internal conflict, as in *Hamlet*?), which points in the plot have chiefly to be brought out, and what the *shape* of the piece is (this dictates variations of pace, of rhythm and of sound level). *Situations* are what matter, not lines or speeches, which are what actors tend to concentrate on. As in the case of talks, few submitted plays are fit to go into rehearsal as written. An experienced producer nearly always has something to add to, or more commonly subtract from, the author's final draft. But creators must be treated with respect. It's monstrous (and, where copyright material is concerned, illegal) for scripts to be hacked about without the author's permission. The wisest course, as with talks scripts, is for a producer to make suggestions, find out if they stand the test of discussion, and leave it to the creator to do the rewriting. Minor tidying up is a different matter.

Before the script is retyped and duplicated, do make sure that

126

the technical directions are as clear as they can be made. This is one way of saving rehearsal time. Directions of another kind, however, are usually best deleted. Some writers bestrew their scripts with adverbial advice to the actors on how to speak their lines – (*angrily*), (*tenderly*) and the like. Nine times out of ten this is unnecessary. Take, for example:

JEREMY: (*hesitantly*) Er, do you think – um – you could possibly lend me ten pounds?

How could such a line be spoken *except* hesitantly? Other writers go in for underlining important words in the text. Unfortunately, for very good reasons, the important word is not always that on which the stress actually falls. So that to underline, when not superfluous, is often misleading.

Now and again, of course, plot or characterisation require a speech to be uttered in a way that is far from obvious. In that case the dramatist is right to provide a hint. But 'interpretation' should in general be worked out during rehearsal, not pre-determined by the author. So try to clear the script of underlining and adverbial advice.

It's at the editing stage that you should take a careful look at the author's ideas about *transitions*. Until fairly recently it was assumed that if listeners were not to be confused there had to be a brief pause (or, very occasionally, a prolonged cross-fade) between one scene and the next. The conventional technique was to fade out scene A, pause, and fade in scene B. Sometimes a variation on the technique might be employed: scene B, for instance, might be started at full volume. But very seldom did one escape the pause. Now everything is speeded up. Inter-scene pauses are still used to signify lapses of time; otherwise producers normally *cut* from scene A to scene B. The gain in fluidity is obvious. Less obvious perhaps is the restrictive effect of per-petually having to distinguish between one scene and the next by textual forecasting or sharp acoustic changes. When these expedients can't be used, or when a moment of rest or reflection is desirable, one can still gratefully fall back on the old formula. Whatever decisions you come to you should see that they all appear, clearly expressed, in the edited version of the script.

For the sake of clarity I'm keeping the various aspects of production separate, but in practice they overlap. You'll have

127

formed your conception of the structure and character of the piece at a very early stage, so you'll have already settled on your *style* of production. The sounds you want your actors to make will have started to float through your head: in other words you'll have made preliminary decisions on casting and acoustics. All these thoughts have now to be translated into detailed practical terms. Let's take an example. We've mentioned the difficulty of indicating physical action through dialogue. Radio dramatists in their text and 'stage directions' often provide no more than a figured bass, a groundwork of suggestions. It's your business to bring the writing to life. Begin by deciding what *precisely* is meant to be happening, moment by moment: only when you have comprehensively envisaged the action can you hope to make it clear to the audience. Suppose that two or three characters are engaged in conversation when 'an angry crowd approaches'. In my experience the average author after writing in

(*Distant shouts and yells*)

then lets his characters talk on until the mob is practically on top of them. This is thoroughly unnatural. When do the characters start to notice the uproar? A little later than we, the radio audience, because they are concentrating on their conversation. But notice they must, and react they must, long before the crowd actually comes round the corner. It is for the producer to fill in the author's sketch with additional lines and carefully-contrived pauses. Then the approach will carry conviction on the air. Non-visualising radio-drama producers, of whom there are not a few, will broadcast the scene as written – alas.

I shall be saying later that rehearsals ought to be journeys of exploration. But this doesn't imply that a producer's study of the script and initial planning can ever be dispensed with. The more careful the preparatory work the more fruitful the exploration.

CASTING

This is the next step, and a crucial one. Casting, someone said, is sixty per cent of any production. A business-like attitude always helps. If you have many performers to choose from their names ought to be in a card-index. Each card should carry, in the case of new artists, a summary of the audition report, and in other cases your comments on successive performances (together with notes on any unexpected skills revealed during rehearsal, as when Irene had to read Sandra's lines and did rather better than Sandra). A producer who depends on his memory is unfair to the members of a desperately over-crowded profession. What is more, his plays will all sound the same.

In radio you cast for voice, not appearance. This may seem a glimpse of the obvious, but it's a principle often forgotten. That fine radio actor Felix Felton was condemned for years to play fat men, merely because he was himself so fat. And yet, with his superb vocal control he could *sound* as thin as anyone. What a waste of talent!

Another point. In the theatre one usually casts on an individual basis: such-and-such a player for such-and-such a part. In radio the over-all pattern of sound is nearly as important as the individual voices. Again, since listeners must be able to distinguish readily between one character and another, if there's no difference in accent, age or sex there must be a difference in pitch and timbre. So quite often you cast for vocal contrast. (As Lefeaux puts it, what one then needs is not the voice beautiful but the voice peculiar.)

In general it's unwise to expect a performer to sustain an unnatural voice for a long time. Occasionally (as when an actress is required to play a small boy) there's no alternative, but it's always a tricky business: the vocal mask tends to fall off unexpectedly.

What about accents and dialects? The artist David Jones once told me that some academic draughtsmen could invariably catch a likeness while others couldn't, and that their abilities in this direction bore little relation to their academic competence. Similarly, certain actors can reproduce an accent more or less at will, while others, equally good in other respects, make a fearful hash of it. In radio, *authenticity* of accent and dialect is vital, and actors in the second category ought to be reserved for straight parts and the few accents they can manage. The likelihood is that among any body of listeners there will always be some who are familiar with the accent being assumed; they'll be sadly put off by a Nigerian masquerading as a Kenyan or a

Glaswegian as a Tynesider. Even listeners who don't know a particular accent can still *feel* that an actor is having trouble with it: to that extent the play will cease to convince. After all, we are thinking in terms of sound.

The final rule I would suggest may seem strange but makes sense: *cast for co-operation*. In radio the selfish actor is a menace. This is clear enough in the context of a crowd scene. One's cast is always too small, and if some of those present save their voices by 'goldfishing' instead of shouting the whole scene will collapse. More to the point, one has to remember that *radio acting, like radio listening, is primarily an affair of the imagination.* The actor has no scenery, no properties, no costume, no make-up, to get him into the right mood: he depends entirely on his own imaginative powers – plus the emotional help given him by the rest of the cast. I've often been aware during a production that the actors and actresses sitting around were willing, almost forcing, their colleague at the microphone to succeed. When this atmosphere, this link of temporary but intense feeling, doesn't exist, all you get is a succession of people addressing empty air. When it does exist you get genuine interaction; the actors play *to each other* rather than to the microphone; every speech has a heightened significance; the setting seems to take shape before your eyes and the whole play develops a new force. So steer clear of the performer who reads a newspaper while the man at the microphone is giving his all. Radio drama depends on teamwork.

ENTER THE STUDIO MANAGERS

Sooner or later you must tell your technical crew *in detail* what you want by way of sound-effects, music, acoustics, reverberation and so forth. Once upon a time every producer would meet his crew a few days before first rehearsal, explain his ideas, and give the technical arrangements a trial run. Now BBC studio managers have to guess what's required. In consequence, no sooner do rehearsals begin than they are interrupted, actors get stopped in mid-flight, and exasperation sets in among the cast as crew and producer argue amongst themselves. Not a good system!

The place of music in plays we shall discuss in the next chapter.

We've already referred to *sound-effects*; but here are some additional thoughts about them.

Not all effects are meant to be realistic. They can be employed to indicate what a character *thinks* he's hearing (in a house-breaking scene they might well be 'unnaturally' loud). Radiophonic sounds are particularly useful for the heightening of tension. You should distinguish between the various uses to which effects can be put, and aim always at consistency of style within a given production. To slide from naturalistic, or objective, effects to subjective, or impressionistic, ones and back again calls for the most delicate management.

Naturalistic effects themselves, particularly such standbys as doors and footsteps, should also be used as consistently as possible. If a door is heard to open, not heard to close, and then heard to open once more, listener-bewilderment can rapidly set in.

Odd though it now seems, until a few years ago indoor scenes were regularly played against a backcloth of studio silence. Only now and again, e.g. when the script called for a character to exit through a french window, were we allowed to hear sounds from outside – passing traffic, usually. It was H. B. Fortuin who first realised the artificiality of this convention. Since we never get complete silence in real life, Fortuin decided that his drawing-room and kitchen scenes should include haphazard (or apparently haphazard) effects – barking dogs, traffic, footsteps, wind – at very low level. To these noises from the outside he would add indoor sounds – a clock chiming, for instance – which again bore no apparent relationship to the dialogue. The result was a startling increase in verisimilitude.

Subconsciously perceived and irregularly introduced sound effects have another use. Even though the setting of a particular scene has been clearly established, *listeners need to be reminded of it* if the dialogue continues for more than a few minutes. Otherwise they start to forget where the scene is taking place: the voices come from a sort of limbo. You can get the author to slip reminders into the text, but as a rule effects do the job more neatly. Take the instance of a historical play in which conspirators were heard plotting late at night in, of all places, a church-yard. The scene went on and on. All one needed, to conjure up the setting, was, *very occasionally*, the hoot of an owl and the crunch of a heel on the gravel path.

A.E.R., OR FOLDBACK

In most drama studios it's now possible for actors to hear (at low level) the recorded effects which are being played behind their dialogue. By means of a technique called A.E.R. (Acoustic Effects Reproduction), or foldback, the output of the tape and/or disc machines in the cubicle can be split on its way to the mixing panel and one stream diverted into a studio loudspeaker. A.E.R. can be extraordinarily helpful to an actor. It's all very well to know in theory that he's delivering his lines against the roar of Atlantic breakers and howling wind. It's quite another thing to *hear* the recorded gale, even faintly. With A.E.R., scenes played against noisy backgrounds are always far more realistic, because the actors adopt a more appropriate vocal quality. And they time their lines better, relatively to the sound-effect, than when they depend on light-cues.

 The acoustic of a recorded effect may differ from that of the studio voice associated with it. Or the perspectives may not match: the effect may, for instance, have been recorded in close-up while the actor is required to be in the middle distance. A.E.R. can bring voice and recorded effect into a much closer aural connection. If the A.E.R. loudspeaker be placed near the actor and its output picked up by the actor's microphone, the total sound-effect as heard by listeners is very satisfactorily 'coloured'.

Sometimes an external source of sound is supposed to be in a changing spatial relationship to the characters (as when a car drives right up to them). If you want a lifelike result it's not enough to start the recorded effect very low and fade it up. Get your studio managers to 'A.E.R. it' and, as the car get closer, to mix in more and more direct sound from disc or tape. This will alter the acoustic, and therefore the apparent perspective, in a most realistic way.

It must be admitted that A.E.R. is difficult to handle. The positioning and the level of an A.E.R. loudspeaker are both critical. Unwanted 'echo' is an ever-present danger. Nevertheless it can give results attainable in no other way.

SPOT EFFECTS

These are the sounds made in a studio without the aid of tape or disc – real footsteps, slammings of doors, and so on. Most producers, and most studio managers come to that, don't worry much about them. Usually they are left to the least experienced member of the crew. This can be a mistake, for poor spot effects can destroy a play. In radio they are not mere 'noises off' but an integral part of the production, another strand in the pattern of sound. *'Door closes'* looks simple enough on the page. But *how* is the door to be closed? Is the character striding out of the room in a fury? If so the door needs to be banged. Or is he creeping out? If so it must be closed quietly and carefully. And how *soon* after the line should we hear the door? Points like these go to the heart of a play's credibility.

CO-ORDINATED EFFECTS

If your playwright's ideas are to be fully realised you shouldn't think of recorded effects and spot effects as separate. Co-ordinate them as artistically as you can. Let's take a very simple instance. The setting is a house in the tropics. There is a distant earthquake (for which we have been prepared in the text). No doubt you could find a recorded rumble to indicate the earthquake, and no doubt that would do well enough. But how much sharper the impression if you add the spot effect of a glass shaking on the table.

PERSPECTIVES

It's painful to see, as one sometimes does, a spot effects operator stuck with all his impedimenta in one corner of the studio while the cast mill round the rest of it. However smoothly the panel SM mixes the microphones, if the spot effects and the dialogue come from opposite ends of the room the result can't possibly be convincing. When a character is supposed to lift a telephone off its cradle and talk into it, the instrument (and the effects operator) ought to be actually at his elbow: if they aren't, the perspective of the effect will almost certainly be wrong.

In any scene all the speakers, plus the sound effects, should be in a recognisable spatial relationship to one another, *and to the listener, who is the fixed point*. Take a courtroom scene. If the listener is presumed to be among the spectators he can't simultaneously be on the bench with the judge, so the judge's voice must come from a slight distance. The defendant, though, and the cross-examining barrister, will sound rather nearer. Alternatively, if the listener is supposed to be with the defendant, the latter will be in aural close-up, the judge will sound fairly near, and the cross-examining barrister quite far away. And so on. You, the producer, know which character in a given scene you are chiefly interested in. As a rule you put him nearest the microphone (which represents the listener). The other characters you place at varying distances from it, until your sound-picture is complete. Always remember that when a character is well removed from the aural fixed point he should *not* be heard distinctly. If the judge on the bench and the defendant in the box and the policeman at the back of the courtroom are all equally audible there is something wrong.

Perspectives, once achieved, don't have to be static, as though you were doing a Purcell opera. In real life people move about as they talk. Whenever possible (i.e. whenever their positions are not acoustically critical) you should, within reason, let your actors move similarly. Continual small changes in the angles at which they address the microphone make the sound more interesting and more lifelike. Individual performances also improve: actors in straitjackets can't give of their best.

Major changes in position, as when a character crosses a room, are often indicated by footsteps, followed by the use of the voice in a different relationship to the microphone, thus:

HENRY: (*at a distance*) Why Lucy! How nice to see you.

FOOTSTEPS APPROACH AND STOP.

HENRY: (*near*) Where have you been all this time?

But if the character is made to move *as he speaks* the varying acoustic will in itself make everything clear.

Incidentally, you can often produce a pleasant impression by getting an actor to move (as he speaks) out of the range of one microphone and into that of another – e.g. from an indoor acoustic to an outdoor.

It's sometimes possible to alter your sound-picture *during* a scene. In a courtroom drama, for instance, if prosecuting counsel, standing at a distance from the listener, were pressing

134

a vital point, you might make the microphone seem to zoom up to him like a television camera. But to make this kind of change without confusing the listener is much more difficult than might be supposed.

To have the sound-picture in your head is one thing; to achieve it quite another. The equation is complicated by so many variables, ranging from the carrying-power of actors' voices to the technical characteristics of microphone and studios. Sometimes, if your characters stand or sit or lie in approximately the positions they would occupy in real life they *sound* right. At other times, for no apparent reason, they don't. You have then to propel your actors round the studio until, after a process of trial and error, every character sounds as though he's where he ought to be. The process isn't much fun for the actors, so remember to explain what you're at and to ask for their co-operation.

ACOUSTICS

With rare exceptions, the action of a radio play occurs in a series of recognisable settings. Sometimes these call for sound-effects – the twittering of birds, the creak of rigging, or whatever. And they always call for (but don't always get) the appropriate *acoustic background*.

Voices in a small, heavily-furnished, carpeted room sound quite different from voices in a large, bare room. In an old cathedral they reverberate endlessly under the high roof and in all those cavernous recesses, while being simultaneously reflected by the hard surface of marble and polished wood. A train, a closed car, an open car, an open field – each has its own acoustic properties. It's the job of your senior SM to simulate these, and many others, as you and the dramatist ordain. In the studios of today, built for aural flexibility, it can always be done – by using the right kind of microphone, by altering its characteristics, by placing it properly, by screening it, and so on (for a fuller discussion of these points see Chapter 14). What the studio manager must have is a knowledge of the laws of sound, a willingness to experiment and a good deal of patience, since variations in temperature and humidity can radically alter the acoustics of a studio, so that an arrangement of microphones

135

and screens that worked perfectly yesterday may not work at all today.

Studio managers like the rest of us are prone to take the will for the deed. I've seen three microphones put up in different parts of a studio to yield, not, as intended, three different acoustics but the same acoustic thrice repeated, so that the cast were kept in a state of constant locomotion to no purpose whatever. It's for the producer to close his eyes, *listen* for acoustic differences and, if they are not apparent, insist on getting them.

The adroit employment of changing acoustics often does more to excite a listener's imagination than any number of 'effects'. And the stimulus is the more valuable in that for the most part it's received subconsciously.

MANAGING THE MANAGERS

A producer should work closely with his studio managers. They are his adjutants, and he can't hope for a successful recording if he's at cross-purposes with them. Often they possess, in addition to expertise, a store of artistic sensibility which it's wise to tap. But *you* must take the final decisions. The division of labour is simple: as producer it's *your* business to know exactly what sounds you want: it's *their* business to get them for you, in their own way.

When a recording is about to begin it's usual (in the BBC at any rate) for the panel operator, who is the senior SM, to address the cast over talk-back – 'Stand by: good luck: we're going ahead in ten seconds from . . . NOW.' At other times he should talk to them through you: 'Would you mind asking X to move nearer the mike?' If you merely chew a pencil while your panel operator tells the actors where to stand and how to move you'll look as though you've abdicated, and they'll lose confidence in you.

It's important, on the other hand, to maintain the authority of the senior SM vis-à-vis his colleagues. If you give instructions to the tape or spot operator direct you will only cause confusion – plus a certain glum resentment in the breast of the panel operator which will do your production no good.

Studio managers, like actors, prefer not to be directed in

detail. But radio programmes, so delicate is the medium, are in fact accumulations of detail lovingly assembled. On the stage if a trumpet-call 'off' occurs two seconds late nobody worries: the actors can fill in with 'business'. In radio such a gap would wreck the play. So if a disc or tape operator persists in fading up a particular sound too slowly (or too quickly) you must do something about it. If your panel operator gives you a fraction too much artificial reverberation, do something about it. Don't carry interference to the extent of being a nuisance to people who have their hands full; don't 'conduct' every fade; but if some detail is wrong, speak up.

THE PLAYERS

One of the big moments of any production comes when you meet the assembled actors. If they are professionals they will be slightly on edge. So should you be. You may not think much of this particular play. Never mind. To professional actors *every* engagement is important, and they expect a professional producer to take a similar view.

How do you start off? After the (highly important) introductions and hullos some directors launch into a kind of seminar, or, worse still, lecture, on the play's significance and the style in which it should be performed. Others, with whom I humbly associate myself, prefer a more down-to-earth approach. I tell the actors what they need to know about the mechanics of the play – in particular, which sequences take place at which microphones – they should be numbered – and exactly where the light cues occur. (It usually takes an hour or so to ladle out this information – at dictation speed, so that the actors can pencil it into their scripts.) Then we start *a complete run-through*, with music (if any) and effects.

Needless to say, during this run-through, or stumble-through, every mistake that can be made is made. Actors go to the wrong microphones and have to be redirected over talk-back. 'Effects' are brought in too early or too late, or missed out altogether. But we press on, stopping only when something gets hopelessly out of phase. And by the end we shall have achieved some serviceable results. The actors will know that the script works: they will have confidence in it. I shall know to what extent they share my basic conception of the play. I shall have probably picked up several ideas that were better than my own. On a more pedestrian level I shall know which sequences are going to be awkward technically, and I can make a shrewd guess at the overall duration. Quite a solid basis on which to start detailed work. What, one may ask, will the 'seminar' have achieved? At worst the producer will have irritated his actors by telling them what they know already. At best they will have gained some purely intellectual grasp of the play. In the case of a particularly opaque text the process may be useful, but H. B. Fortuin's remark sticks in the memory, 'OK, they understand the meaning. So what?'

Some producers begin by reading the play through. I wonder why. Unless it is so technically demanding that it will have to be rehearsed and recorded in separate bits, the read-through is a pointless exercise – a hangover, probably, from the leisurely routine of the theatre. Nothing really *happens* until the actors position themselves at the microphones and start doing their stuff in earnest. This is when a play begins to exist. And only when it exists can those involved in it – actors, technical crew, producer – begin to feel its full significance, or significances.

The rehearsal process is a journey of exploration; a journey, however, of which you are the leader, alternately discussing and deciding which way to go.

RUNNING THE REHEARSALS

Radio's first law is that rehearsal time is always inadequate. So you must make the best use of what there is. *Plan* your rehearsals; know in advance how long you'll spend on each stage. You won't stick to the plan precisely but it should save you from disasters like having to record with the last few pages unrehearsed.

If you run into a technical hold-up, do explain the situation to your cast. Never leave them in suspense, wondering when the rehearsal will be resumed: this builds up quite the wrong sort of tension. If you know that some of them won't be needed for a long time, tell them so and let them go off to the canteen if they want to. In a word, treat your people with consideration.

Actors and actresses are sensitive, high-spirited, endlessly charming. But many of them are insecure personalities; practically all of them live from hand to mouth. This should be borne in mind if they show off, or make nuisances of themselves in other ways. Remember, you and I might not be as balanced and rational as we are if we had to spend half our lives on the dole.

Never treat your performers merely as a collectivity – 'the cast'. If you do you can abandon hope of a successful production. Your job is like a jockey's, with the difference that you have to take not one but a whole string of horses over Becher's Brook. Each of these creatures, often so beautiful to look at, demands separate treatment – a word of encouragement here, a tug at the reins there, a pat, the prospect of a lump of sugar, a light flick of the whip, as the case may be.

It's a great thing to know when you've got all you can out of an artist. Few sights are more painful than that of a producer nagging away at an actor who is simply not capable of doing any better. When he's reached his limit the only sensible course is to say you're satisfied. Otherwise the wretched man will go to the microphone haunted by self-doubt and give a final performance which is worse than ever. Production is a struggle, and as in other forms of struggle truth is often the first casualty. (Not that you should be *too* reckless in your misstatements. It's a mistake to distribute insincere compliments like trading stamps.) Genuine approval is a different matter. It's far too rarely expressed. Acting can be a strain on the whole personality to a degree few non-actors realise. A good actor literally gives part of himself to every role. At the end of a difficult scene he is surely entitled to some more inspiriting comment than 'Right; let's break for lunch.'

NOTES

At intervals you'll stop the rehearsal for 'notes' – comments on a run-through of the play or a particular sequence. Never try to give notes over talk-back. Even if all the cast can hear you, no discussion, no meeting of minds, is possible. Go into the studio. And *take your senior SM with you*. If the technical side of the operation is to work in smooth co-ordination with the acting he must be privy to your *thinking*, and the cast's. And there's the further point that any new decisions you make may

139

require him to change his technical arrangements. So the sooner he's in the picture the better. If you alter the mechanics of the play (effects, light cues, moves) see that *everyone* in the cast remarks his script. A change that doesn't concern Mr A today may do so tomorrow.

Incidentally, in the case of a new actor you should make sure that he knows *how* to mark a radio script. The main point is that marks should be *easily seen* and *precise*. A light cue (often indicated by an asterisk or by F, for flick, followed by the number of the microphone) should be pencilled in *immediately* before the first word to be uttered, thus:

DISTANT SHOT.
3. CAPTAIN: *3 What's that?

Cuts should be indicated not by half-hearted squiggles but by a line, or lines, drawn firmly from the last word left in to the next word left in.

An actor is less likely to miss a cue if he *underlines* the name of his part every time it occurs (I mean in the margin), thus:

3. CAPTAIN: *3 What's that?
4. SOLDIER: It's coming from the harbour, sir.

Notes on interpretation should be as specific as you can make them. And they should be positive. To ask, as I've heard a young producer do, 'Could you get a bit more life into it?' is not only insulting but useless. The kind of direction an actor values is, 'When she says "Tomorrow" you realise you've been tricked. So shouldn't you speak in a different tone from page nine, line twelve?' Incidentally, when an actor agrees to alter his reading of a particular speech, do make him try out the new version at the microphone. If you don't, you may well find during the recording that all unconsciously he reverts to his original interpretation. Or, in his anxiety to please the producer, he may go overboard in the new direction.

A producer often has to be highly specific when arranging for 'reaction'. Some performers, given a brief to react ad lib, do so with such enthusiasm that those addressing them get completely thrown. In such cases you have to decide the exact points at which reaction should occur – and, sometimes, what exactly should be said. It's much the same with crowd scenes and

140

'struggles', both of which are apt to sound unrealistic and confusing. People making meaningless noises *sound* like people making meaningless noises. It's worth taking the trouble to find appropriate words and phrases, to be shouted on pre-arranged cues.

Timing, which basically means distributing the pauses within a line or a speech, is one of the foundations of radio acting. Should you ever suggest an improved phrasing, or, come to that, an improved inflection? In other words should you ever direct a radio actor on minutiae? The conventional answer is no. Actors usually resent it: it suggests they don't know their jobs. Furthermore a few actors have such bad ears that they couldn't reproduce an inflection if they tried. And actors are supposed to build up their performances from within. This is entirely true; but in radio there's not always time for ideal solutions. So short cuts – 'try saying it like *this*' – occasionally become inevitable. Of course you have to be very sure of your ground before you give *pointilliste* direction. But radio, as we've seen, is a *pointilliste* medium, an affair of shades and nuances. The words are in close focus, and untidy phrasing or false inflections, which the best actors are sometimes guilty of, really hurt.

TEMPI AND DYNAMICS

On the stage, visual interest can, and often has to, compensate for vocal monotony. In radio, as we know, variety in the sound pattern is indispensable. So when giving your notes make sure, as a conductor does, that the tempi within the piece are well contrasted. You can arrange for differences between the pace of one scene and the next or between the tempo of dialogue and narration. Or you can get a character to talk faster or slower than those he is addressing (this can be difficult to achieve, though, since actors tend to fall gradually into a *uniform* tempo, especially if they are reading mechanically and not concentrating on the sense: when it happens you should prod them pretty sharply). As for variations in the *volume* of sound, fortissimo and pianissimo are both denied you by electronic necessity, but you can work wonders by *sudden* changes from forte to piano, as when one line is shouted from the middle distance and the

141

next murmured in close-up. In a single-sense medium well-judged contrast is all.

CUTTING

Every producer hates having to shorten a play for reasons that have nothing to do with art – simply to force it into a given time-slot. But one often has no choice. You can, of course, postpone the moment of decision by recording the play as it stands and cutting it later in the editing channel. The trouble with this technique is that the raw material for your razor blade is by then unalterable. You may find that in order to make one cut you must, if you are to be logical, make several more: thus you end up with an *under*-run. It's wiser to perform major surgery in the studio, where instant rewriting can often hide the wounds. What you should aim at, I suggest, is a recorded production that's *very slightly* too long. Every production has its unhappy details which it's pleasant to be able to pare away in the channel.

To cut a script in rehearsal needs caution. What may be an 'easy cut' from one point of view, necessitating no change in the mechanics of production and no rewriting, may do monstrous damage from another. Your first responsibility is to be fair to the play and the playwright.

Actors, provided their own roles aren't affected, prefer a few big cuts to a multitude of tiny ones. To lose two or three pages is not difficult: to lose a lot of half-speeches and interjections can throw an actor, especially if the cuts are made late, when he's beginning to have his lines by heart. The need to look out for pencilled marks on the script drags him back to the initial stage of getting to know a play from the outside, when he's ready and anxious to work from within.

There's another kind of script change, the emendation. If you find that a line as written doesn't come off, it's right to alter it. But try to do so early on. After a certain point it's best to follow the advice I gave talks producers and leave the text alone.

GET BACK!

Once your notes and the subsequent discussions are over, get back, I beseech you, to the cubicle where you belong. Drama producers are not paid to be mere *répétiteurs*, concerned with taking the cast through their lines: they are responsible for the transmission *as a whole*, and this can be properly heard, and reacted to, only in the cubicle.

This is when you should remember a point made in an earlier chapter, that radio audiences have no scripts. Do try to put yourself in their place now and again: listen without looking at your own script – except to make hasty marks against passages in need of attention. You're almost certain to find that various sound-effects, which seemed quite all right when the script told you what they were *meant* to convey, in fact convey no information whatever. Or that words and phrases, heard without being simultaneously read, suddenly seem less than satisfactory. Even your tempi, your pauses, may now seem subtly wrong. If it's important for talks producers to devote some of their rehearsal time to listening without looking, it's infinitely more important for drama producers, in their far more complex role.

'STALENESS'

Two final points about the running of rehearsals. First, do remember that whereas you have been living with the script for weeks your cast have had only a few days to familiarise themselves with it, and perhaps only a few hours at the microphone. So even if you're more or less satisfied with their performances and have nothing more to give them by way of direction *don't hesitate to make them go through it all again*. The 'shop stewards', of whom there are one or two in most casts, will take you aside over lunch and murmur their fears of becoming stale. But how can actors become stale in a day and a half? As for the other bogey, tiredness, have no fear: they'll pull all the stops out for the recording, however many times they've been rehearsed. At the risk of sounding brutal I would say that if a cast start the recording a bit tired, a bit hungry, it doesn't matter: it will give an edge to their performance. To let them 'go on' complacent and yet slightly insecure is a far greater risk. Secondly, don't expect your actors to be perfect *before* the recording begins. 'The object of rehearsal is to establish a blueprint for a superb performance.'

RECORDING

As far as the actors are concerned, recording provides the moment of truth. So how you record matters almost as much as how you rehearse. For some years the practice of recording a bit at a time has been much in vogue. Admittedly, some radio-dramatic works are so technically difficult that discontinuous recording, to use the more elegant term, can't be avoided. And some are so episodic that there's no point in avoiding it. And, of course, *occasional* breaks, as at the end of a heavy scene, may actually be welcome to a cast. But producers should always remember that when a piece has real dramatic development, real flow, when its characters grow and change, when the dialogue develops its own momentum, abrupt halts for purely technical reasons can be sheer cruelty to actors. Still worse in these circumstances is the practice of recording scenes, or parts of scenes, out of order. And the ultimate in undesirability is attained when a cast is forced to record every sequence several times over, not because retakes are necessary but so that the producer may later choose a sliver from this and a slice from that to form a radio play which is a collocation of performances no one ever gave.

All this, it may be said, is exactly what happens in the film world. True, but film-makers have inherited, and are apparently stuck with, their extraordinary system of endless retakes of minuscule performances by obedient mimics. In radio we still expect an actor to act, to think and feel himself into his part, to see it as indeed a part of a symbiotic whole. I believe you should be wary of any recording technique that subordinates actors and actresses either to dead machinery or to the godlike wisdom of the producer. . . . But this, I fancy, is where we came in.

POSTSCRIPT ON EDITING

The amount of editing your tapes will need depends to some extend on how they were recorded. Obviously the shorter the takes the more stitching together your technician will have to do. Furthermore, the shorter the takes the more conscious your players must have been of the artificiality of the proceedings; therefore, in all probability, the more they fluffed. So the more slips of the tongue you'll now have to cut out and the more revised versions you'll have to patch in.

This is the boring part of tape editing. It becomes engrossing when you start to put a gloss on your show by eliminating badly-spoken lines which you failed to correct at the time, reducing the level of not very apt sound-effects, sharpening fades and adjusting the duration of pauses. The danger lies in taking the process too far. 'To gild refinéd gold, to paint the lily' is, we are informed, a mistake.

POSTSCRIPT ON STEREO

We haven't referred to stereo so far. The problems of producing plays in stereophony are artistic rather than technical. In fact the technical side of the operation can be very easily mastered: in the BBC a two-day training course is considered quite enough. For stereo productions the studio floor has to be marked out in a series of concentric semicircles, so that actors required to move from right to left or vice versa can execute, for electronic reasons, a series of crab-like shuffles round the microphone. The producer has to plan every position and movement in advance: in order to save rehearsal time he includes these details in the script. (Most of the action these days occurs on a *part* of the 'sound stage' only, usually the middle. Too much vocal ping-ponging, as in early stereo, makes the play sound ridiculous.) Once they know their positions, actors (and spot-effects operators) use a stereo microphone in much the same way as a monophonic one; there's no reason, says Raymond Raikes (and who should know better?), for rehearsals and recordings to take much longer than they would in mono.

Tape editing is certainly more difficult in stereo, but the channel is a place of opportunity. By 'spreading' or 'panning', monophonic sound-effects can be made stereophonic or moved about the 'sound stage'. By copying and reversing the tape, characters can be moved from one side of the 'stage' to the other. Most mistakes can be rectified.

What about the artistic problems? Stereo brings with it many advantages, of course. Radio drama is all sound, and stereo sound is more lifelike than mono, which 'cannot reproduce the spatial and directional effects which are part of everyday aural experience'. Monophony can tell you whether a sound comes from near or far: it can't tell you whether it comes from your right or your left or from in front. Stereo can. Furthermore (I am quoting from a BBC pamphlet), whereas in mono 'sound sources originally separated in space, and picked up by no matter how many microphones, are inextricably mixed together, finally to emerge from the loudspeaker as a single stream of sound', stereo provides *aural separation*. In other words stereo enables you to cope with, and distinguish between, several sounds going on simultaneously, just as you can in real life (provided your two ears are all right) and just as you can't when you listen to mono. What a gift to the producer of a long scene played over music or effects!

No one doubts that the extra 'information' afforded by stereo adds a new dimension of richness to broadcast *music*. And in the case of music that's all that matters, since to most varieties of it radio is merely 'a means of transportation' (in George MacBeth's phrase). But radio drama is in a different category. Radio drama is an art-form, and a switch from mono to stereo implies an alteration of the terms in which it communicates with its audience.

I remember the first impact of sound on the hitherto silent world of cinema. Most of the fluidity and pace which films had acquired disappeared overnight, and we seemed to be back in the era of the photographed stage play. Some people hold that the advent of stereo has had a comparably cramping effect on radio drama. The flexibility of construction which is one of its main attractions largely depends on the scope it offers for very brief scenes and extremely rapid transitions. In stereo the fade-out and the cut are alike difficult to manage. Often the best a producer can do is to airlift his characters from one side of the sound stage to the other, making the change more acceptable by panning an intervening sound-effect. Obviously this isn't an operation to be repeated too frequently in the same play. So authors have to write longer scenes and directors are reduced to using incidental music as inter-scene punctuation – back to the Thirties with a vengeance.

Another charge made against stereo drama is that every character, and indeed every sound-effect, has to be *somewhere*. So we return to the days of the Blind Man's Theatre, separated from the sound stage (significant phrase) by an invisible proscenium arch. The buttonholing monologist who is everywhere

146

and nowhere (a key figure in classical radio drama) can't exist in stereo without electronic fudging. The quintessential radio play was defined by Martin Esslin as 'not so much a play you can't see as one you don't need to see'. He might have added 'or couldn't possibly see'. This concept seems far removed from some current productions in stereo – though not, fortunately, from all. Various leading producers, John Tydeman for instance, seem to be reacting against the new/old staginess. A number of recent stereo plays, excellently if not very fluidly constructed, have shown that the weight of extra 'information' need not prohibit treks into interesting subjectivist territory.

The trouble is that the stereo audience remains minute. The BBC has been transmitting drama in stereo since 1958: we have been given many ingenious adaptations of stage successes, and some exciting original pieces by writers like James Saunders, Peter Redgrove and Rhys Adrian. But the ever-increasing number of stereo productions are usually heard on mono sets. This is a doubly unhappy situation: not only do most listeners miss out on the stereo; they now get a much poorer brand of mono than they used to (for 'compatibility' see Chapter 14).

What opens up an exhilarating but highly uncertain prospect is Quadrophony. Before the listener can be profitably placed in the very middle of the action we shall require a quite new kind of dramatic writing. If the scripts of quality are not forthcoming, however, quad drama, as a tarted-up version of conventional drama, will presumably remain a mere oddity, one of those 'absurd essays in literalism' motivated by the misguided belief that it makes a dramatic experience more 'real' (to quote Philip French's description of a very similar technique being tried out in the cinema).

12: MUSIC IN PLAYS AND FEATURES

All radio consists of speech, music and pure sound. Perhaps we should add a fourth ingredient, silence. From one point of view the whole art of radio production is no more than the use and combination of these elements, the old alliance of music and dramatic speech being particularly fruitful.

In this alliance speech commonly takes the lead. But there have been some interesting attempts to make the partnership more equal. 'The Dark Tower', already referred to, was a landmark in that the action of the play was based on musical themes. Later, in Alan Sharp's 'The Long-Distance Piano Player', which was a psychological study of a man attempting to set up a record in non-stop playing, the music (by Richard Rodney Bennett) was obviously pivotal. A few other productions, like the prize-winning 'Ballad of Peckham Rye', should be mentioned with respect. On a more popular level we've had all those dramatised biographies of composers, conductors and opera-singers in which the text served chiefly as a setting for the musical gems (I remember producing one such programme which, in addition to fifteen actors, had six solo singers, three choirs, an orchestra and a brass band). And out of the mists of memory come those old-time radio revues and concert party shows in which music and 'book' were on a par. Stage-derived they may have been, but their occasional brilliance indicated a line of musico-dramatic development which ought to have been followed up. Alas that it never was.

The notion of radio opera as the ideal synthesis of music and drama has, on the other hand, been actively encouraged; but for some reason no amount of talent, money or publicity could make it come to the boil.

In the area of documentary and narrative features the attempt to use music as creatively as speech seems, after a hopeful start, to have run into the ground. We have mentioned the sad eclipse

148

of Charles Parker's Radio Ballads. Charles Chilton now makes his distinctive narrative features – popular history based on popular song – less often. A few years ago Michael Mason did give us a feature ('Rus', a celebration of Mother Russia) in which an almost ceaseless barrage of stereo music and 'effects' was improved and reinforced by multi-track recording, radiophony and every variety of electronic distortion and reverberation. But this spirited if unsubtle attempt to fire off all the guns in the armoury of sound has, to the best of my knowledge, had no successor.

What it comes to is that in contemporary plays and features music continues to play merely a supporting role. But it's a role of immense potential, worth far more attention than it usually gets.

OBJECTIVE AND SUBJECTIVE

One can use music, like sound-effects, objectively or subjectively. 'Objective music' is the kind that goes into a documentary because it exists in the area of life being reported on, or into a play because it's an integral part of the scene (as when characters are talking in a discothèque). 'Subjective music' (background music, incidental music, mood music) goes in to create or heighten an atmosphere. Artistically (and, in the case of features, factually) it's of the first importance that the two categories should not be 'promiscuously confounded'. This is not to say that they can't be used in the same programme. The music of 'The Long-Distance Piano Player' changed from the totally objective at the beginning to the almost totally subjective at the end, when it matched the hysteria of the protagonist; but exercises of this kind call for skill and experience.

Subjective music can give a tremendous lift to a production; increase the emotional charge. It can suggest a period or a place. It can offer oblique and witty comment on the action. Or, in the old-fashioned way, it can provide inter-scene punctuation or modulation – the so-called music link. But it can do more harm than good. Music links excessively prolonged become separations, interrupting the flow of the drama. Background music generally can be too familiar, inappropriate in scale, or, in the case of stereo productions, too real and solid.

149

The early producers plastered music all over their plays. And, judging by what I remember of their broadcasts, their knowledge of the repertoire was confined to the more whistleable passages of Tchaikovsky and Sibelius, together with Elgar's 'Enigma' Variations, which did valiant service over many years. A reaction then set in, and during the next period music was practically barred from radio plays and dramatised features unless it had been specially composed. Practice in the BBC has now changed again. Budgets have become so tight that most producers can no longer afford to commission original music. At the same time they have grown weary of austerity. The wheel has come full circle and we are back in the Thirties and Forties, with music on gramophone records being freely employed, but fortunately with more taste and discretion than of old.

SOME POINTS OF DETAIL

To make a suitable choice of background music is not enough. The way it's handled counts for just as much. Difficult though it is to lift short passages out of a much longer work, it's possible to do the job gracefully. The secret is to *follow the grain* of the music – to introduce it not just anywhere but at the beginning of a phrase, and to end (this is even more important) when the music itself comes to an end, momentary or final. (If you simply can't find a decent exit-point you can always 'lose' the music by means of a slow fade behind the next bout of speech, which provides a necessary distraction; but this trick should be a last resort.) What you should avoid is cutting off slices of music according to the clock. Producers who say 'I want twenty-five seconds of music here' or 'Let the disc run on to the next minute' are musical butchers, whose activities can be guaranteed to offend any listener with an ear.

Another point. It's not necessary for the music to be always faded in and faded out. If the bits are carefully chosen you can often start or end them 'full', which is normally more effective.

The sustained use of *music behind speech* presents problems, especially in mono. (It should go without saying that what we are discussing is *instrumental* music behind speech. To have a voice speaking one set of words while another voice sings a different

set is on the whole a rotten idea.) You can make things easier
for yourself if you choose legato rather than staccato passages.
Beware of 'bumpy music whose peaks don't relate to the natural
rhythms of the words'. In adjusting the respective levels of music
and speech you need a nice judgment. You should never let the
music overpower the words: listeners find this infuriating. On
the other hand there's no point in having background music if
it's only a faint buzz.

You can often make a sequence smoother if you get your
panel operator to 'play the fader' rather than, as so many operators do, holding the music at a fixed modulation throughout.
In my experience it usually pays to 'establish' the music first, to
dip it fairly sharply (at a predetermined, musically suitable,
point) as the speech begins, and to bring it up equally definitely
during speech breaks. Needless to say, for this you need an
operator with a feeling for music and a quick, confident hand
at the controls. When the speech consists of narration, why not
get the reader to work with headphones and imperceptibly to
'spread', or close up, phrases so that his speech rhythms tie in
with the rhythms of the music? Provided you have plenty of
rehearsal time it's possible in this way to achieve a most elegant
and pleasing effect.

SOURCES OF 'SUBJECTIVE MUSIC'

Apart from ordinary gramophone records there are albums of
'mood music' on sale. I have used some of these well-orches-
trated snippets on occasion: very handy they are.

It's unfortunate, but a great deal of the standard orchestral
repertoire always evokes visions of dinner jackets and lady
harpists in black silk dresses. In any case, the spread of musical
knowledge has made it more and more difficult to lift bits from
classical works: so many listeners can tell where the bits come
from, and are irritated when they get cut short. *Radiophony* can
provide us with 'music uncluttered by association' or, as another
expert puts it, 'a background of semi-abstract sound, on to
which the listener can project his own images of God, nature –
anything'. Radiophony is a combination of two art-forms. The
first is *musique concrète*, the electronic treatment of sounds

151

existing in nature or in 'conventional' music. The second is (completely) *electronic music*, the manufacture of which is a laboratory process at every stage. It was in the Fifties and Sixties that the possibility of using the science of electronics to generate sound textures which had never before existed began to excite composers, mostly on the Continent. But English musicians like Tristram Cary have made outstanding contributions in this area, and the BBC's Radiophonic Workshop has now been in business for a good many years.

How long radiophony in drama can remain 'uncluttered by association' is anybody's guess. It seems to me that it's rapidly building up a set of associations of its own – with 'Dr Who', with horror films, with grotesquerie of every kind. But there's no doubt that for the producer it's an extremely useful adjunct to conventional or, as some would say, real music – and to naturalistic 'effects': many practitioners have decided that its greatest value for them is as a cornucopia of sound-effects, heightened, modified or totally imaginary.

It's worth pointing out, for the benefit of producers in developing countries, that you don't need expensive equipment to get perfectly respectable radiophonic results. Near-miracles have been achieved by studio managers who employed only oscillators, ordinary tape machines (for, e.g., speed changes, reverse playing, tape feed-back and delay) and perfectly ordinary reverberation and distortion. What you must have, though, is a willingness to experiment and a fair amount of time.

No doubt composers for the media, as for the concert hall, will increasingly offer us fusions of radiophonic with non-radiophonic music. In the meantime the average producer's best bet (provided he has the money to lay it) is still specially-commissioned 'conventional' music. It needn't cost the earth. A very few instruments in unexpected association can give admirable results (Antony Hopkins' background music for the Nesta Pain features remains a classic instance of this truth). The use of *single* instruments is a further possibility which has been only very occasionally explored (but then usually with success). To quote an example from TV, the 'Dixon of Dock Green' signature tune, which underwent so many transmogrifications, never sounded more interesting than in its original form – for solo mouth-organ.

152

WHAT ABOUT THE DRUMMERS?

Drama producers in developing countries, African countries particularly, are in one respect luckier than they realise. In their societies music remains a thriving folk-art, and the fees demanded by its practitioners remain absurdly low. It seems to me that the reluctance of so many African producers to build, drama-wise, on this living foundation is a waste of a great opportunity. So often one hears African radio plays on traditional local themes, or on timeless universal themes, adorned with European music which is inappropriate either in general or to the rhythms and overtones of the actors' voices. Successful attempts have been made to create truly African operas and musical plays for the stage. Why isn't more done in radio?

And then there's the art of improvised song. I've already re-ferred to a visit to Enugu where I saw an Ibo-language play being rehearsed and recorded. It wasn't only the cast who improvised: after each scene a group of local singers and instru-mentalists, sitting on the studio floor, would provide a spon-taneous comment on what had passed and an introduction to what was to happen next (the endlessly drawn-out diminuendos of the drummers were particularly thrilling). Music of this kind, used with imagination, is surely capable of adding a new dimension to radio plays in other parts of the Continent.

13: OUTSIDE BROADCASTS

I

ADAM SMITH RIDES AGAIN

Until very recently radio OBs seemed to be heading for extinction. I remember taking the whole subject off the syllabus of BBC Staff Training as no longer worth general study. Outside Broadcasts seemed the clearest instance of the box's ability to beat the set: who, given the choice, would listen to a description of an event in preference to seeing it for himself, with expert commentary thrown in? The argument, as far as it went, was unassailable. What it didn't take into account was the adaptability and enterprise which, under conditions of fair competition, people are apt to show when their livelihoods are at stake.

The BBC's radio OB staff have grabbed every chance to exploit the difficulties of television in the U.K. – the cumbrousness of its equipment, its heavy operating costs (swollen by restrictive practices on a scale unknown to radio), the shortage of alternative channels, and the consequent rigidity of its programme planning. Though radio still provides set-piece OBs for ever-diminishing audiences, its OB men have concentrated more and more on turning out supplementary material, much of it specialised, using the profusion of wave-lengths at their disposal – broadcasts of local and regional occasions, unheralded visits to a whole succession of sporting events (the multiple OB), all-day cricket commentaries, and so forth. The public enjoys this kind of material, and British television at its present stage of development is on the whole not geared to supplying it. So the BBC's radio OB department remains very much in business, and the laws of classical economics are once more vindicated.

Of course there are many parts of the world where TV coverage is still limited. In these countries set-piece radio OBs, particularly ceremonial ones, continue to be important.

154

OB producers are modest men, prone to regard themselves as primarily managers and fixers. Certainly they usually have a tangle of practical and business requirements to sort out before a major OB can occur. Various bodies have to grant their various permissions and to be kept sweet throughout. Commentators have to be found, along with engineers, to service the operation. Post Office lines, mobile equipment, the programme budget – all these can present problems.

For an OB point in frequent use there tends to be a standard microphone lay-out. In other cases the producer, in company with an engineer, should visit the site in advance and decide on a feasible arrangement. I say 'feasible' rather than 'the best' because often the interest of the producer and of those organising the event don't coincide. Take a religious service. To you as producer the obvious place for one of the mikes may be right in front of the pulpit. But if it obscures the congregation's view of the preacher you'll have to put it somewhere else. Rarely can you insist on getting your own way, and even more rarely is the struggle worthwhile.

Your visit will be even more rewarding if you take your commentator along. He'll do his job twice as well on the day if he's satisfied with the commentary position. The ideal is that he should be comfortable – how would *you* like to commentate with nowhere to rest your notes or with one leg of your chair over the edge of the rostrum? – and able to see those aspects of the event that matter most.

An instance of what can happen when the site isn't properly examined sticks in my memory. It was long ago, in the early days of the BBC's Welsh Region. King George VI and his Queen were doing a post-accession tour of Britain, and it fell to me, under the direction of a very young Wynford Vaughan Thomas, to describe their ceremonial entry into the Guildhall at Swansea. Wynford and I arrived, in carefree style, a short time before the royal couple, and I took my place, as instructed, just inside the main doorway. Five minutes later an outraged official pointed out that if I stayed there I should be commentating within two feet of His Majesty's left ear. This was undeniably true. There was a broom cupboard near by. Abandoning the

Town Clerk, the High Sheriff, the two Assize judges and the rest of the reception party, I stepped inside, microphone in hand, and there I stayed, behind a firmly closed door. By the light of an unshaded bulb I concentrated partly on re-reading my notes and partly on keeping still, so as not to cause an unseemly disturbance by knocking down some twenty dustpans which were in a precarious pile just behind me. As soon as I heard on headphones that Their Majesties had begun to mount the Guildhall steps I counted ten, took a deep breath, and launched into a description of what I sincerely hoped the royal visitors were doing. I ended, I remember, with a flowery coda, composed for me in advance by Wynford, in which I bade farewell to Their Majesties as they disappeared into the Main Hall. Then I stepped out of the cupboard, to find them still there. . . . It was an unnerving experience, which, given less luck, might have brought two promising radio careers to a premature conclusion. Certainly it impressed on me the necessity for careful reconnaissance of the site by all concerned.

LIAISON

Promoters of ceremonial occasions, which occur so often in the Third World, normally organise them fairly well ahead. The producer of the OB should make it his business to attend the planning meetings, so that his point of view may be taken into account from the beginning. To come in at a late stage and ask the Committee to alter their arrangements for the sake of the broadcast is to ensure maximum ill-will.

Incidentally, far too many producers expect their engineering colleagues to be thought-readers. When you're building up to a big OB, keep them in the picture. Don't be content with conversations in the corridor: circulate plenty of paper (why should administrators have a monopoly of it?) giving details of what's being proposed (and of alterations to the proposals). And do consider if on the day itself everyone wouldn't be happier and more secure if he had *a typed running-order*, listing microphone positions, timings, the sequence of commentaries, handover cues and so forth.

In simple OBs such as ten-minute visits to football matches a commentator can be his own producer – and indeed his own panel operator (not too difficult when all he has to do is plug a suitcase-full of equipment into a permanent line). Complex OBs in which many microphones are brought into play demand an OB van, with mixer, talk-back facilities and a producer in charge. Talk-back is essential: you, sitting in the van, *must* be able to communicate at any moment with your commentators, all of whom should be wearing headphones, or earpieces. Try to give your instructions clearly, concisely, and (to repeat an earlier suggestion) when the recipient is not himself on the air. Of course, if all you want to do is direct someone to start (or stop) commentating it's simpler and more businesslike to give him a light cue. He in turn should be able to signal *you* when he wants to slip in a bit of unplanned commentary: if he can't attract your attention quickly a lot of vivid stuff is liable to be lost.

An OB producer is more than a fixer. Admittedly his raw material is fluid, and largely outside his control, but the *way* it's presented depends very much on him. Many of his decisions are bound to be unpremeditated, but they are production decisions for all that. And it's important that what he says goes, whether commentators and engineers like it or not.

What makes a good radio OB? Primarily suspense, the feeling that the listener is assisting at an event that's happening *now* and working up to a real conclusion. This is as true of a State Opening of Parliament as of a football match. It follows that although the recording of OBs for subsequent transmission is sometimes unavoidable a recorded OB seldom has much impact.

II

COMMENTARIES

The running commentary is a highly specialised form of solo broadcasting, but it's a form that many radio practitioners, particularly in small stations, are compelled by circumstances to have a go at. Robert Hudson, formerly the BBC's Head of Radio OBs, says that in modern broadcasting 'the gift of the gab' is the last attribute you need. But you must have a reasonable degree of fluency. If you want to do acceptable radio commentaries it might be a good idea to practice lucid utterance in everyday life and deliberately to extend your working vocabulary, i.e. the stock of words you use, as distinct from those you understand the meaning of. I've heard that excellent broadcaster Raymond Baxter admit that in his early days he made constant efforts in these directions.

In a commentary you start by telling the listener what you can see (and he can't). But gradually, as his imagination becomes stirred by your words and by the 'noises off', he begins to feel that he's at the event with you. Your eyes somehow become his, and you seem to be not so much telling him what's happening as remarking on a scene which lies before both of you.

This idea of a crasis, or union of personalities, dictates a commentator's whole approach. 'I wish you could see the team coming out. They look super – in the pink of condition.' Why is this an appalling bit of commentary? Because of the clichés of course, and the gush, and the generality of the description, which conveys to the sporting listener nothing he wouldn't know or couldn't guess. But above all because it draws unnecessary attention to the fact that the listener isn't present. Compare this: 'Ah – and now the . . . team are running on, Jock Brown as usual three seconds behind the rest. Smith looks absolutely fit – no trace of a limp.' This is better. The phrases carry concrete information and the speaker clearly knows what he's talking about. More materially, what he's saying is exactly what an interested spectator might say to himself or the person sitting next to him. The commentator's role, I suggest, is to sustain and amplify the sort of monologue which most of us go in for when we watch a game or a great ceremony.

'BE PREPARED'

Knowledge, expertise, authority – whatever you may call it, this quality is essential to the commentator. The moment you say 'I think this must be the President's car' or refer to 'one of the forwards' instead of 'Charlie Jones', your effectiveness starts to sag. The union of personalities depends on your being smoothly omniscient. So preparation is vital. If you intend to commentate on a game, watch the players in action beforehand, so that on the day you can recognise every one of them at a glance. If your assignment is to describe a ceremony, sit through the rehearsal, so that you'll know *precisely*, not approximately, what is supposed to happen.

As important as physical is *mental* preparation, by which one means talking to the organisers of a ceremony, reading handouts and mugging up playing-records. On the day there may be a hold-up at any moment. It's elementary that you *explain* why the proceedings have come to a stop and then 'fill in' with apt remarks. But these won't come out of the top of your head. You can only meet the challenge if you have background information galore.

Most BBC commentators like to make written notes, finding that the act of putting details on paper helps the memorising process. But whether notes can be used in the OB itself is another question. For fast-moving games they are as good as useless; they can, however, be a great standby in ceremonial events. Write your notes on stiff cards, so that they won't rustle or get blown away. And *phase* them – so many cards for so many divisions of the ceremony. When you've broadcast the facts on a card, drop it on the floor to make sure that you don't repeat yourself. 'Reading what's on the card' has bridged many an awkward gap. But you should never be hooked on this kind of help. When something goes wrong you need to be able to make the felicitous remark *instantly*, not after consulting your notes. To have the background information literally at your finger-tips is good: to have most of it in your head is better. And I shouldn't like you to think that background knowledge is valuable only in crises. Your entire commentary should sparkle with it. In this respect as in others Richard Dimbleby was a model. On the State occasions he specialised in he constantly illuminated the formality of his subject with sudden shafts of incidental information, usually of a personal sort. Other commentators at the ceremony of Trooping the Colour might say: 'The Colour-Sergeant is marching forward'. Dimbleby would refer to 'the Colour-Sergeant (whose last Trooping this is – he leaves the Army next week)'. The ability to flavour a commentary with pinches of 'associative material' of just the right kind at just the right moment is a sure sign of the master practitioner.

But of course prepared material must take second place to what you can see. And what you see includes not merely the central features of an event but any number of incidental occurrences. Highlighted, these often bring an otherwise predictable commentary to sudden life. So don't ignore the dog scampering across the parade ground: work him smoothly into your description. 'Smith is going to throw in' may be a perfectly accurate bit of commentary, but 'Smith, just wiping the mud off his face, is going to throw in' makes a more arresting picture. Good commentators fall like hungry sparrows on the smallest crumbs of human interest.

Colour has a strong emotional appeal. But you need to be specific about it. There's no percentage in 'the beautiful tints of the great East window' or 'brilliant hues' or 'a colourful procession'. By contrast 'The President's wife is dressed in white, apart from an enormous lime-green hat' means something.

You will naturally aim at being in key with the event you're describing, and adapting your style to the occasion. BBC commentators, to my mind, are a little too uniformly bland (as BBC interviewers used to be). I wish they would utter some of their devastating remarks during OBs instead of after them. However, that's only a personal opinion.

When an event is dramatic the excitement will be apparent in your voice and your choice of language. One shouldn't make a deliberate effort to point it up – unless one is assisting at a merely frivolous affair. 'It's terrific!' or 'What a fantastic sight!' are phrases that would hardly do for a State Opening of Parliament, but they would be in order at most film premières.

THE WHOLE PICTURE

It's a safe rule of thumb that you should begin a commentary by rapidly sketching in *the whole picture* (in the case of a cricket match, what the ground looks like, how much of a crowd there is, how many small boys you can see sucking lemonade through straws, and so on). If the event has already started you should also bring the audience up to date at the earliest possible moment – in this case by giving them the state of play.

The whole picture will make sense only if you establish your own position. At a rugby match you might say 'I'm looking across the half-way line, with London Welsh on my left. The ball has just gone into touch, just outside the Rosslyn Park 25 on the far side of the field.' Having fixed your own position you should thereafter relate everything to that. As Robert Hudson says, 'far' and 'near', 'right' and 'left' are infinitely more useful as reference points than 'the South Stand' or 'the Pavilion end'. You can then get down (quickly or at your leisure as the case demands) to a description of the action. But the setting shouldn't be forgotten. Some commentators, after giving a prepared piece about it at the beginning of the broadcast, devote themselves entirely to the nitty-gritty. This is a mistake. However riveting the action may be, occasional glances at your surroundings always give your commentary a lift.

KEEPING IN STEP

In some OBs, soccer matches for instance, the commentator is expected to go practically non-stop (even so, he can't describe everything, and he usually keeps a few moments behind the run of the play precisely in order to be selective). In slower-moving games like cricket frequent pauses are entirely acceptable, especially when they fill themselves up with the evocative sound of bat striking ball or gentle summery applause. In ceremonial broadcasts the sounds of the event itself – the music, the shouted words of command, the noise of marching and counter-marching – can so stir a listener's imagination that it becomes a crime to talk across them. It's a wise commentator who knows when to be silent. A skilful commentator too: the OB man who is not in step with the event, who has to break off for, or still worse, goes on talking through, the National Anthem or the first words of the Speech from the Throne, is merely advertising his incompetence.

14: TECHNICAL NOTES

Most of the generation now making programmes for the BBC and Independent Local Radio have played around with microphones and tape recorders since they were at school. But in developing countries technically-oriented programme staff are still rare. These notes are intended for programme people who, like me, will never grasp Ohm's Law but who realise that artistic advance in radio usually proceeds from a technical starting-point, and who are not content to be clay in the hands of the engineers.

The basic theory of sound is best explained by analogy. When you throw a stone into a pond you create ripples, concentric ripples which spread out towards the bank, thus:

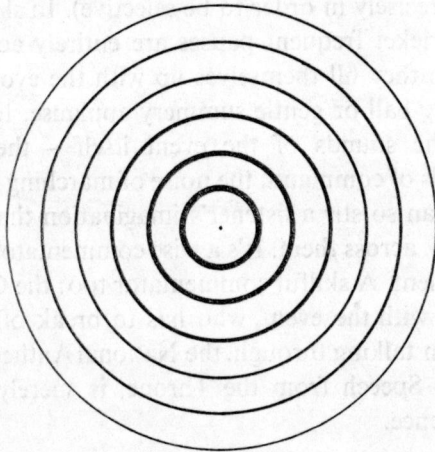

The further away from the point of impact the fainter the ripple. When you make a sound you set off similar concentric ripples in the *air*. In order to speak you have to vibrate your vocal chords; the air vibrates in sympathy. If there's someone near

162

you, those vibrations, those *sound waves*, reach him and cause his inner ear to vibrate: it's this last set of vibrations, translated into electrical impulses, which act on his brain and make him hear you.

So what does a *microphone* do? Like an ear, it accepts and registers sound waves. It also converts them into electrical impulses. For the purposes of broadcasting these impulses are *amplified*, put on to a *carrier wave*, and sprayed in various directions from a transmitting *aerial*. When you tune a receiving set to the same *frequency*, or rate of vibration, as the carrier wave it picks up these signals, *turns them back again into sound waves*, and sends them out into the air of your sitting-room via a loud-speaker. (In a portable set, receiving aerial and loud-speaker are concealed within the box.) Diagrammatically the sequence can be represented thus:

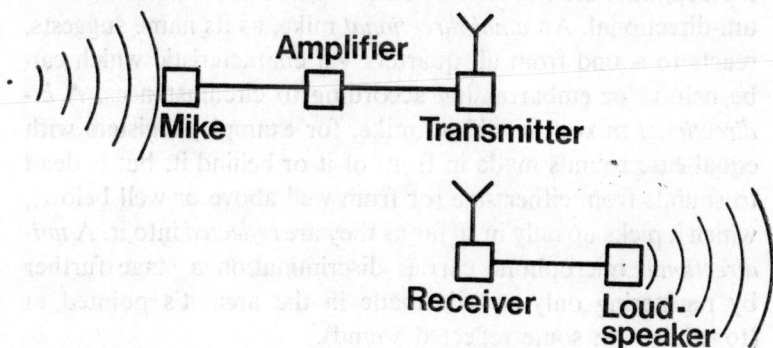

ACOUSTICS AND BALANCE

Now let's look at a studio. When an actor or speaker says something, the sound, we know, flows outward in a series of waves. If the studio walls are hard and bare the waves will bounce back, knock into waves going in the opposite direction, and be *reflected* at various angles and high speed. This process is called *reverberation* (or, wrongly, echo). The barer the room the longer the reverberation period.

Contrariwise, if the studio is filled with soft, absorbent surfaces – curtains, carpets, wall-coverings – the reverberation period will be exceedingly brief.

163

A drama studio is usually divided into two parts, one reverberant ('live'), the other much less so. The less reverberant – the so-called 'dead' – part provides the acoustic background for open-air scenes; the other half does well for scenes in baronial halls or empty houses.

Out of this fundamental acoustic contrast between 'live' and 'dead', producers and studio managers can contrive a whole lot of further adjustments and refinements by moving the microphone, by altering the electronic characteristics of the microphone, by screening the microphone or by varying the position of a performer in relation to the microphone. This is the delicate art of Balance.

TYPES OF MICROPHONES

Microphones are classified as omni-directional, bi-directional or uni-directional. An *omni-directional* mike, as its name suggests, reacts to sound from all quarters – a characteristic which can be helpful or embarrassing according to circumstances. A *bi-directional* mike (the ribbon mike, for example) registers with equal ease sounds made in front of it or behind it, but is dead to sounds from either side (or from well above or well below), which it picks up only in so far as they are *reflected* into it. A *uni-directional* microphone carries discrimination a stage further by registering only sounds made in the area it's pointed at (together with some reflected sound).

The field, or angle, over which a microphone picks up *direct* sound is represented by its *polar diagram*. The basic polar diagrams are:

(*omni-directional*),

Mike

(*bi-directional*, or figure of eight) and

Mike

(*uni-directional*, or cardioid, so called because the polar diagram looks vaguely like a heart).

Some microphones have fixed polar diagrams. Others, particularly various kinds of condenser mikes, are electronically switchable from one polar diagram to another – an extremely useful facility.

DIRECT AND REFLECTED SOUND

In some ways reflected sound is nearly as important as direct sound. All music depends for full richness of effect on some degree of reverberation, which is why ribbon microphones continue to be used for every kind of music except pop. Properly positioned, a single ribbon mike can cope with a full orchestra, the side facing the players registering mainly direct sound and the reverse side reflected sound. Pop music producers, for various reasons, normally opt for a deadish acoustic and a battery of cardioids, each aimed so closely at a single performer (or single section) as to pick up only direct sound. The brilliance is added later, artificially.

For radio drama you normally need a fairly lively studio, not only to give full value to the music of the voices but also to establish an aural frame of reference and to convey impressions of *movement* – impressions which a listener receives only when the *ratio of direct to reflected sound* in an actor's utterance varies

165

as he alters his position vis-à-vis the microphone. So radio-drama producers remain faithful to bi-directional microphones, with cardioids employed now and again, e.g. in unduly reverberant studios. Omni-directional mikes are no good for creating sound pictures and therefore not much good for drama. (An incidental, but very real, advantage of the 'figure of eight' mike over the cardioid is that, because the back is as sensitive as the front, actors in a dialogue sequence can stand opposite each other, and look at each other, instead of having to face the same way.)

If you start with a lively studio you may need to damp it down – perhaps you have only one speaker in a space big enough for twenty, or perhaps you want to create an open-air acoustic. This is where *screens* come in handy – provided you know how to use them, which not all producers, or even studio managers, do. Most screens have a soft, or absorbent, surface on one side and a hard, or reflecting, surface on the other. In the case of a single speaker, a single screen, used in conjunction with a bi-directional mike, can reduce reverberation to a surprising extent. The *hard* side of the screen should be at an angle to the microphone (see opposite). The sound waves which pass above and below the microphone will be reflected by the screen on to wall 'B'; from that they will be reflected on to the 'deaf' side of the microphone, and largely lost.

But not entirely lost. If reverberation remains excessive, or if you want to group several people at the microphone, you'll need more screens. In extreme cases, especially if you move the microphone away from the studio walls, you may have to pretty well surround it with screens.

Remember that nine times out of ten (and contrary to the opinion of many studio managers) it pays to turn a few (not all) of the *hard* surfaces inwards, towards the microphone. If you turn all the soft sides inwards, what you'll get at best is an almost entire absence of reverberation, which is a most uncomfortable acoustic to speak in. But the soft material on a screen is usually tacked on to the framework with no intervening air-space, an arrangement which tends for various reasons to absorb only the higher frequencies, the top notes. So what you'll get at worst is a boomy, bassy acoustic – thoroughly nasty. (Fibreglass, once used for the treatment of studio walls, produces a similar effect.)

166

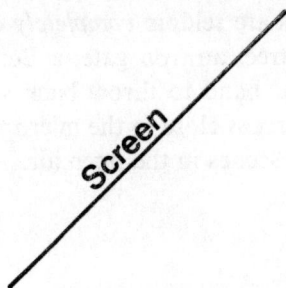

Screen

Studio Wall 'B'

Voice

At a pinch you can hang blankets over the screens (hard sides turned inwards) in such a way that blankets and framework don't meet: that gives you the necessary air-spaces. If your studio equipment includes an R.S.A. (Response Selection Amplifier) the studio manager can help by electronically reducing the number of lower frequencies that reach the console. But this is a fairly drastic sort of operation which ought not to be undertaken without thought, and certainly not as a first resort or short cut.

Two reminders. Remember, first, that hard screens round a microphone should be arranged *irregularly*. If you place them directly opposite one another you'll defeat your own efforts by starting off a whole new series of reflections. Secondly, remember that in real-life open-air acoustics are seldom *completely* dead. There's almost always a wall, a tree, an iron gate, a Bedouin tent – something or other near at hand to throw back sound waves. So one or two reflecting screens close to the microphone can sometimes be useful even for scenes in the open air.

ADDING REVERBERATION

If you need *more* reverberation there are various means of getting it. The combination of a single reflecting screen and a bi-directional mike can be applied to brightening an acoustic as well as damping it down. This time you put the microphone in the middle of the studio and the screen *squarely* behind the microphone, thus:

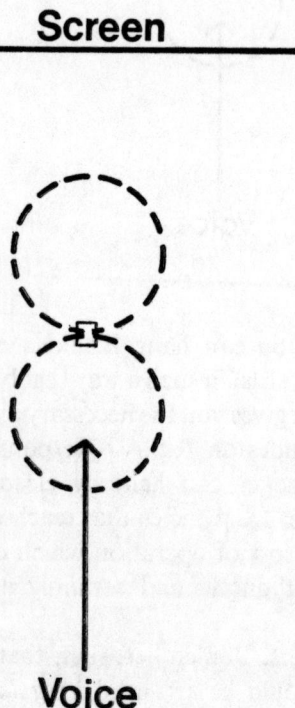

Screen

Voice

The sound waves passing above and below the microphone, instead of going on to the studio wall and (since the studio is by definition deadish) being partially absorbed, will bounce smartly off the hard screen and into the back of the microphone: hence an increase in reflected sound.

In a large studio you can get extra reverberation out of an *atmosphere mike*. The trick is to place an auxiliary mike in such a remote part of the studio that it registers reflected sound only; then, via the console, you mix in to the output of the main mike as much reflected sound as you choose.

Remember that in real life there's less reverberation on quiet speech than on loud. The disadvantage of the old-fashioned 'echo room' (a disadvantage it shares with more modern devices such as 'plate echo' and 'sprung echo', which provide a time-lag variable by electronic means) is that it adds the *same* amount of reverberation to *every* sound. The atmosphere mike, on the other hand, because of its distance from the sound-source, fails to pick up the quieter passages. Hence a more natural effect.

Even if you decide to use echo room, plate echo or sprung echo it often pays to connect these contraptions to an atmosphere mike rather than the main mike.

SIMPLE BALANCES

A producer should be able to balance a speaker or two without the intervention of a studio manager. Start by *listening* carefully. Sit directly in front of the cubicle loudspeaker and fairly close to it. If you don't, the odds are that the sounds you hear will be coloured by reflections from the walls of the cubicle itself. Don't turn your loudspeaker up too high: the volume shouldn't be much greater than that of a home radio set. In balancing, as in the ordinary routine of production, it's a good idea to listen without looking. As I've said before, once you start watching the performers your eye begins to take over from your ear: you no longer make judgements purely in terms of what you're hearing.

When balancing only one speaker at a cardioid or a bi-directional mike make sure that he faces it fair and square.

Make sure also that his mouth is on a level with the business-end of the microphone – not, as sometime happens, several inches higher or lower. And don't let him hold the script between the microphone and his face! Elementary errors, but they can cause a lot of trouble.

Unless – an unlikely contingency – you want maximum reverberation, don't put the microphone anywhere along the axes of a rectangular studio. In this diagram the dotted lines show the positions to be *avoided*.

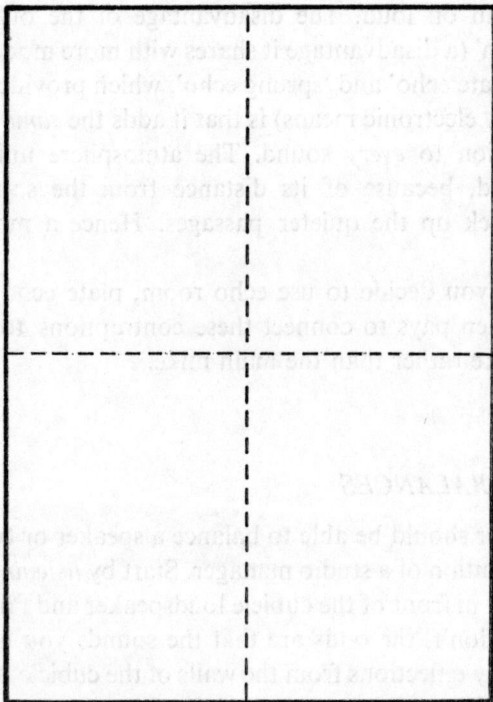

Another bad position for the microphone is near the window between studio and cubicle. Reflections from the sheet of glass will colour the sound and guarantee you a very unattractive acoustic. If you must have your speaker near the window – e.g. in order to give him hand cues – at least see to it that the mike and the window are not in parallel. This position:

| Wall | Window | Wall |

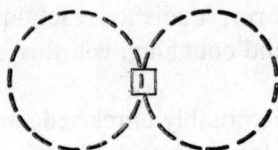

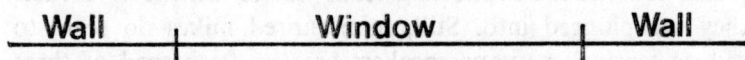

will give dreadful results. This one:

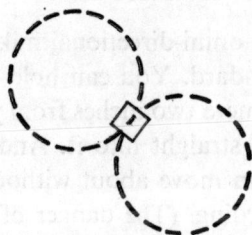

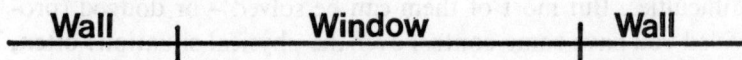

| Wall | Window | Wall |

will just about do. A reflecting screen *behind* the performer will improve matters further.

If you have several speakers to balance, e.g. for a discussion or brains trust, it's possible to use two ribbon mikes (though separation is hard to achieve) or an omni-directional mike (but this tends to pick up a lot of unwanted 'information' from other parts of the studio). A good solution, usually, is to suspend a cardioid mike over the table, the business-end pointing downwards and all the participants sitting round it. That way all the voices get equal treatment.

For an audience show it often pays to use a cardioid mike

pointing at the platform; 'audience reaction', together with the inevitable shuffling and coughing, will then register as reflected sound only.

If your speakers are notably unrelaxed, lanyard mikes (omni-directional and hung from the neck by a piece of string) or other miniaturised mikes are useful. It must be admitted that they are liable to get mixed up with the wearers' clothing so that speakers suddenly become less than audible. And that incautious move-ments can wrench connecting leads out of whichever sockets they are plugged into. Still, miniaturised mikes do seem to reduce tension: nervous speakers are less frightened of them than of the ordinary variety, which they can *see* stuck in front of their faces.

INTERVIEWING ON LOCATION

For this purpose the omni-directional mike in one or other of its forms is now standard. You can hold it as far away or as close as you like – a mere two inches from your face, if you talk across it rather than straight into it. And it's a tough micro-phone, which you can move about without causing too many crackles on the recording. (The danger of crackles from loose connections, by the way, is reduced if, when holding the mike, you wind the connecting lead once round your wrist.)

Outside the studio you come across all kinds of acoustic difficulties. But most of them can be solved – or dodged (pro-vided you have some control over the physical situation: often, as with most news interviews, you haven't, in which case you can only hope for the best).

The sort of room in which interviews tend to occur – the typical large office, for example – is usually far too lively. If you can open windows and doors, do so: this introduces extra absorption. Moving to a corner of the room is normally a good idea (provided you don't sit facing a wall). Drawing the curtains is another good idea, especially if they are thick and heavy. But the best method of coping with excess reverberation is simply to reduce the gain on your recording machine and *work close* to the microphone. This is also the way to deal with unwanted background noise, which, unless masked as I've suggested, *will always sound louder on a mono recording than it did at the time.*

Never, if you can help it, hold your microphone over a hard-topped table or desk. The extra reflections will do the speech quality no good at all.

How should you and your victim range yourselves? It's usually best to sit alongside each other, as though occupying adjoining seats at the cinema. If you sit opposite him the distance between your face and his will be considerable, so that for close working you'll have to wave the mike about, pointing it now at yourself and now at him. Though you often see this done on television it's a most undesirable practice. It slows the interview down (which may not matter on TV but matters a lot on radio) and perpetually alters the balance (ditto). Most of all, it puts the interviewee off: nobody likes having a microphone thrust at him like an offensive weapon. Sitting alongside each other, half looking at each other and half looking down at the microphone, you and he can concentrate on what's being *said*, and the mike can be held perfectly steady – unless your victim changes his position more than usual, in which case you'll have to move the mike in parallel. (You'll also have to move it if he suddenly drops his voice or starts shouting: *it's no use trying to fiddle with the gain in mid-interview*.)

The recording machine should not be parked on a resonating surface (e.g. a polished table), nor should it be near the microphone. If you forget these precepts you'll find that you've taped not merely an interview but a steady mechanical rumble.

If possible every interview should be preceded by a balance test – a genuine one: the other kind is a waste of time. Don't be content with glancing at the meter while your victim and you say 'Monday, Tuesday, Wednesday, Thursday, Friday, Saturday', or something equally fatuous, in turn. You'll find that when the interview gets going he will use a very different tone of voice. And so will you. The only way to achieve a satisfactory balance is by asking a proper question (usually on an unrelated subject) and getting a proper answer, watching the needle the while. Remember that it's more important for the *interviewee* to register than the interviewer. And that it's better to set the gain too low than too high. The volume of a recording can always be brought up in transmission (at the cost of some increase in tape hiss). But distortion due to over-modulation is incurable.

A little weather-lore. If you're recording in the open and it's a windy day, use a wind-shield. If you haven't got one you can make do with a silk handkerchief or a chiffon scarf tightly wrapped over the microphone head. (A linen handkerchief masks the top frequencies.) If it's raining, remember that the machine and the tape must be kept dry, otherwise the tape will slip and the recording become unusable. If it's extremely sunny, keep your completed recordings in the shade: the danger in this case is that the signal may 'print through'.

You're using *magnetic* tapes. It's easy to wipe them unintentionally by letting them enter a strong magnetic field. To push a couple of boxes of tapes into the glove compartment of your car near the dynamo, or pile them on top of a loudspeaker, or near your own microphone, is quite enough to efface the signal.

When you've finished the interview don't switch the machine off right away: let it run for a minute or so. If the recording has to be edited you'll need slivers from this length of 'silence' – more properly, ambient noise – to disguise the gaps.

STEREO

When stereo broadcasting began (as far back as the Twenties), engineers used a very simple arrangement of microphones: one on the left of the source of sound (usually an orchestra) and one on the right. This worked up to a point, but had the disadvantage that the instruments at either end came over louder and clearer than those near the conductor: there was 'a hole in the middle'. To beat this snag the Americans evolved a system of *spaced microphones*; they added a third, central, mike, its output divided equally between right and left channels. BBC engineers found a different solution. Building on the research work of others, they have devised a single microphone which takes in the whole extent of the 'sound stage'. Actually it's a pair of directional microphones in one casing: by an elegant arrangement the left-hand part of the sound-source is picked up by the left-hand mike, the right-hand part by the right-hand mike, and the middle part by *both*.

The left-hand and right-hand signals used to be broadcast separately, from different transmitters, but they are now 'coded' and sent out in composite, to be divided by the stereo receiver.

Stereo in the home requires only two loudspeakers, not three, so it's clear that the hole in the middle has been plugged at the receiving end as well. How? As follows. Sounds originating on the left of the 'sound stage' are mainly reproduced by the left-hand speaker; those originating on the right by the right-hand speaker – obviously. But sound from the middle of the 'stage' is reproduced *equally* by *both* loudspeakers. This confuses the ears of the listener sufficiently to make him feel that it comes from the area *between* the loudspeakers, i.e. from right in front of him.

Any sound-effect recorded in mono can be fed into one channel, then equally into both channels, and finally into the opposite channel. This techniques gives listeners an impression of *movement* across the sound stage. The ability to 'pan' or 'panpot', a sound (by means of a *pan*oramic *pot*entiometer) is, as I mentioned in an earlier chapter, exceedingly useful to drama producers.

Equally valuable is the 'spreader', a device by which a sound-effect recorded in mono – wind, thunder, rain or whatever – can be electronically extended so as to cover the whole sound stage.

Most stereo transmissions are supposed to be *compatible*, i.e. intelligible to listeners who receive them on mono sets. But the compatibility is often more theoretical than actual. A producer who wants an equivalent to the placeless narrator of monophonic radio drama may resort to using the stereo microphone out of phase. The result is impressive in stereo but can be inaudible in mono. Again, two simultaneous streams of sound – even two streams of speech – can be distinguished one from the other in stereo but are a mere babble in 'compatible' mono. And further, when the positions which actors in stereo take up left and right of centre have dramatic significance this is lost on the mono listener. So are *moves* between left and right. So is the air-lifting that indicates a scene change.

The problems of compatibility are partly technical and partly artistic. But that goes for most problems in radio. These may, of course, also be administrative, legal, financial, political or moral. Hence the endless fascination of the medium.